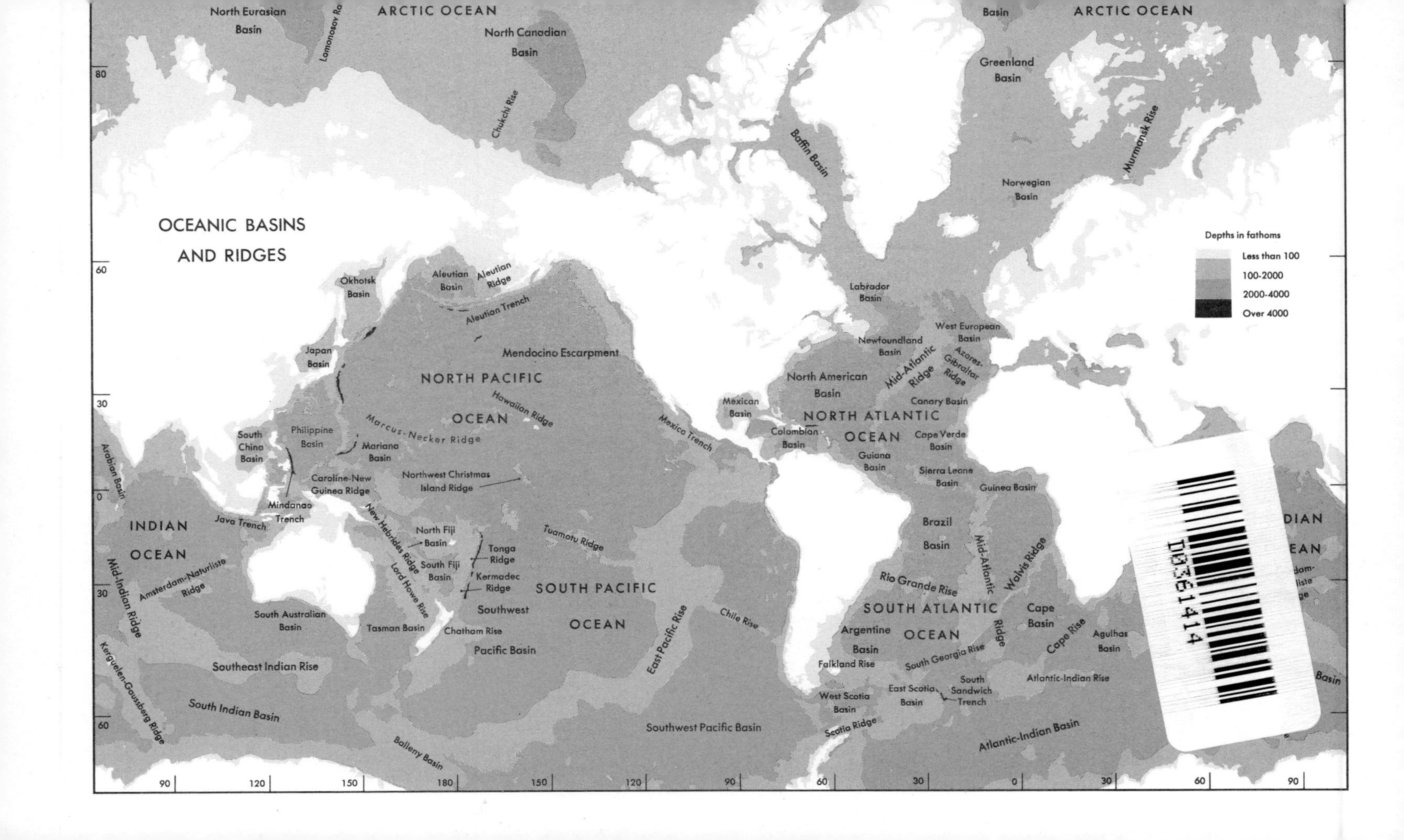

OCEANIC BASINS
AND RIDGES
Depths in fathoms
Less than 100
100-2000
2000-4000
Over 4000
ARCTIC OCEAN
North Eurasian Basin
North Canadian Basin
Chukchi Rise
Greenland Basin
Murmansk Rise
Baffin Basin
Norwegian Basin
Okhotsk Basin
Aleutian Basin
Aleutian Ridge
Aleutian Trench
Japan Basin
Mendocino Escarpment
NORTH PACIFIC
OCEAN
Hawaiian Ridge
Marcus-Necker Ridge
Philippine Basin
South China Basin
Mariana Basin
Caroline-New Guinea Ridge
Northwest Christmas Island Ridge
Mindanao Trench
Java Trench
Arabian Basin
INDIAN
OCEAN
Mid-Indian Ridge
Amsterdam-Naturliste Ridge
New Hebrides Ridge
North Fiji Basin
South Fiji Basin
Tonga Ridge
Kermadec Ridge
Lord Howe Rise
Tuamotu Ridge
SOUTH PACIFIC
OCEAN
Southwest Pacific Basin
South Australian Basin
Tasman Basin
Chatham Rise
Southeast Indian Rise
South Indian Basin
Kerguelen-Gaussberg Ridge
Balleny Basin
East Pacific Rise
Chile Rise
Mexico Trench
Mexican Basin
Colombian Basin
Labrador Basin
Newfoundland Basin
West European Basin
Azores-Gibraltar Ridge
Mid-Atlantic Ridge
North American Basin
Canary Basin
NORTH ATLANTIC
OCEAN
Cape Verde Basin
Guiana Basin
Sierra Leone Basin
Guinea Basin
Brazil Basin
Rio Grande Rise
Walvis Ridge
SOUTH ATLANTIC
OCEAN
Cape Basin
Cape Rise
Agulhas Basin
Argentine Basin
Falkland Rise
South Georgia Rise
South Sandwich Trench
East Scotia Basin
West Scotia Basin
Scotia Ridge
Atlantic-Indian Rise
Atlantic-Indian Basin
Southwest Pacific Basin
80
60
30
0
30
60
90
120
150
180
150
120
90
60
30
0
30
60
90

THE PRENTICE-HALL FOUNDATIONS OF EARTH SCIENCE SERIES

A. Lee McAlester, Editor

STRUCTURE OF THE EARTH

S. P. Clark, Jr.

EARTH MATERIALS

W. G. Ernst

THE SURFACE OF THE EARTH

A. L. Bloom

EARTH RESOURCES, 2nd ed.

B. J. Skinner

GEOLOGIC TIME, 2nd ed.

D. L. Eicher

ANCIENT ENVIRONMENTS

L. F. Laporte

THE HISTORY OF THE EARTH'S CRUST*

A. L. McAlester and D. L. Eicher

THE HISTORY OF LIFE

A. L. McAlester

OCEANS, 2nd ed.

K. K. Turekian

MAN AND THE OCEAN

B. J. Skinner and K. K. Turekian

ATMOSPHERES

R. M. Goody and J. C. G. Walker

WEATHER

L. J. Battan

THE SOLAR SYSTEM*

J. A. Wood

*In preparation

oceans

second edition

KARL K. TUREKIAN
Yale University

PRENTICE-HALL, INC., *Englewood Cliffs, New Jersey*

Library of Congress Cataloging in Publication Data

TUREKIAN, KARL K
Oceans.

(Prentice-Hall foundations of earth science series)
Bibliography: p. 143
Includes index.
1. Oceanography. I. Title.
GC16.T8 1976 551.4'6 75-37647
ISBN 0-13-630426-5
ISBN 0-13-630418-4 pbk.

10 9 8 7 6 5 4 3 2 1

Printed in the United States of America

PRENTICE-HALL INTERNATIONAL, INC., *London*
PRENTICE-HALL OF AUSTRALIA PTY. LIMITED, *Sydney*
PRENTICE-HALL OF CANADA, LTD., *Toronto*
PRENTICE-HALL OF INDIA PRIVATE LIMITED, *New Delhi*
PRENTICE-HALL OF JAPAN, INC., *Tokyo*
PRENTICE-HALL OF SOUTHEAST ASIA PTE. LTD., *Singapore*

contents

four

five

six

seven

eight

nine

introduction

Ours is the only planet in the Solar System with sufficient surface water to form oceans. Because 71 percent of the Earth is covered by water, the study of the oceans is fundamental to our understanding of the Earth. The discipline concerned with oceanic processes is called oceanography, but oceanic processes are so numerous and varied that the science of oceanography is usually further subdivided into four main areas: physical oceanography, which deals with the properties of ocean water in motion; chemical oceanography, which is concerned with the chemical reactions occurring in the oceans; biological oceanography, which includes the study of life in the oceans; and geological oceanography, which is concerned with the structure of the ocean bottom and the processes active there. In the following chapters we will be exploring, principally, the geology and chemistry of the oceans, both present and past. To understand these aspects adequately, we will also have to discuss, in less detail, the circulation of the oceans and the role of biological activity in the seas. Before going on to *what* we know of the oceans today, it will be useful to review *why* this knowledge was obtained—that is, practical as well as scientific reasons that have made the study of the oceans so compelling to scientists and explorers alike.

In the first place, the entire history of the Earth is linked closely to the oceans. The oceans receive the burdens of streams derived from the weathering and erosion of the land. This debris, accumulating on the ocean floor as sediments, records the history of geologic events and preserves, as fossils, a representation of the changing life of the seas.

The oceans also regulate major processes occurring on the Earth's surface. They are the primary source of water that reaches the continents as rain and snow, and they contain the largest reservoir of carbon that is involved in the biological cycle. The high heat capacity of water makes the oceans an important regulator of climate, especially in maritime lands. And the movements of ocean currents are of fundamental importance for both marine life and man.

The swarming, diversified life of the sea has been an important source of food for man through the ages. Mollusks, crustaceans, fishes, whales, and seaweed are, today, major resources for large parts of the world. In addition, nonfood products derived from the animals and plants of the sea have various economic uses, ranging from pearls to building materials.

The chemical resources of ocean water of interest to man include fresh water obtained by desalination and elements for industrial products such as magnesium, potassium, bromine, and iodine. Diamond accumulations on some sea coasts, offshore oil and gas, and phosphate and manganese deposited from the ocean itself round out the list of useful products from the sea.

Aside from the chemical and biological products of the sea, the economic utilization of the oceans extends into the areas of communications and transportation. Transoceanic cables are laid on sediments that may be subject to slumping or distortion as the result of earthquakes and other events. These movements of the sediment may break or disrupt the cables causing costly delays and repairs. For this reason, among others, the topography, earthquake activity, and sedimentary properties of the ocean bottom are studied in great detail. Ships sailing on the surface of the ocean provide the most obvious practical use of the sea, but the increased use of submarines for exploratory, military, and possibly commercial purposes have aroused renewed interest in the properties of deep water. Underwater navigation and remote sensing by sonar have become important for military as well as commercial reasons, hence, the variations in physical and chemical properties of ocean water with geography and depth are part of today's intense oceanographic efforts.

For all these practical reasons as well as the perennial scientific quests of the hows and whys, the sea has become an area of intense study, especially since World War II. The range of techniques available, the number of oceanic research ships, and the number of qualified scientists have all increased tremendously in the past 30 years. The findings of these oceanographers as well as those of the early pioneers will be discussed in the following chapters.

an ocean for the earth

The Earth is different from all the other planets in that its surface is virtually covered with liquid water. We call this liquid water the ocean, or oceans, and they are the subject of this book. Before we can proceed to explore the "water planet" intensively we must understand its place in the Solar System and the circumstances that resulted in an ocean for the Earth.

THE EARTH IN THE SOLAR SYSTEM

Our Sun is a star, one of about 100 billion clustered together in a rotating "pinwheel" array we call a galaxy. Galaxies are the modular units of the universe. There is no doubt that at least one star, our Sun, has a planetary system associated with it. There is no reason to assume that planetary systems are not the general case for the stars of the universe, rather than a rarity.

Table 1-1 shows what we have learned about the inhabitants of our Solar System so far. More than 99 percent of the mass of the Solar System is in the Sun, whereas most of the angular momentum (or spin) is associated with the planets revolving around the Sun. The four planets closest to the Sun—Mercury, Venus, Earth, and Mars—have Earthlike sizes and densities and are called the "terrestrial" planets. The four next outermost planets have masses and densities more similar to the largest planet in the Solar System—Jupiter—and are therefore called Jovian planets.

The Jovian planets are made up of hydrogen and helium in about the same proportions as in the Sun. The terrestrial planets are made of metallic iron-nickel and silicate compounds, which we identify on the Earth's surface as rocks. The

Table 1-1 Properties of the Planets

	Sun	Mercury	Venus	Earth	Mars	Jupiter	Saturn	Uranus	Neptune	Pluto
MASS (EARTH = 1)	329,000	0.054	0.81	1*	0.11	314	94	14.4	17.0	0.05 ?
RADIUS (KILOMETERS)	695,000	2,439	6,050	6,370	3,400	71,000	57,000	25,800	22,300	2,900
DENSITY (GRAMS PER CUBIC CENTIMETER)	1.41	5.42	5.25	5.52	3.96	1.33	0.68	1.60	1.65	3 ?
ALBEDO	—	0.06	0.73	0.39	0.26	0.51	0.50	0.66	0.62	—
EFFECTIVE TEMPERATURE (°K)	6,000	616	235	240	220	105	75	50	40	40
SURFACE TEMPERATURE (°K)	—	616	600	300	230	130	—	—	—	—
OBSERVED AND EXPECTED GASES IN ATMOSPHERE	H, He, O, Fe, N, Mg, C, Si, etc.	—	CO_2, N_2, Ar, H_2O, HCl, HF, O_2(?)	N_2, O_2, Ar, CO_2, H_2O, etc.	CO_2, H_2O, N_2, Ar	H_2, CH_4, NH_3	H_2, CH_4 NH_3	H_2, CH_4, NH_3	H_2, CH_4, NH_3	—
DISTANCE FROM THE SUN (A.U.)†	—	0.39	0.72	1	1.52	5.2	9.5	19.2	30.1	—

* The mass of the earth is 5.976×10^{27} grams.

† A.U. = "Astronomic unit," the mean distance of the Earth from the Sun (149.6×10^6 kilometers).

atmospheres of the Jovian planets are dominated by the hydrogen-helium mixture plus compounds like methane and ammonia that would be formed under these reducing chemical conditions. The atmospheres of the terrestrial planets are different in composition and total pressure but none of them is dominated by hydrogen or reduced compounds.

The atmosphere on Mercury must be very small because the small size of the planet and its close proximity to the Sun would prevent the retention of gases. The surface of Mercury resembles the surface of the Moon, also an atmosphere-free planet. The preservation of impact craters on both these bodies from the early eons of the Solar System's history indicates little or no alteration of the crusts of these planets. The Venus atmosphere is almost totally carbon dioxide at a pressure about 100 times the pressure of Earth's total atmosphere. There is virtually no water in the Venus atmosphere and the observation that the surface temperature of the planet is 616°K rules out the existence of ice caps or oceans. In essence Venus is a "dry" planet. Since water must have been a part of the atmosphere at some time to account for the presence of carbon dioxide, as we shall see later, water as a compound must have been destroyed since the planet originally formed. The dissociation of water by the Sun's rays and the ambient high temperature of Venus would result in the evaporation of hydrogen from the planet and the tying up of the released oxygen by the oxidation of iron-bearing rocks exposed at the planet's surface.

The Martian atmosphere, with a pressure of only one-hundredth of the Earth's total atmospheric pressure, also primarily consists of carbon dioxide. The most dramatic features of the Martian landscape are the ice caps at the poles (composed of carbon dioxide and water) and evidence of fluid processes on the planet's surface.

We depend heavily on *meteorites*—objects impacting the Earth from outer space—to give us some clue to the early history of the Solar System, because they represent material in various stages of accumulation and reconstitution of the primitive components of our Solar System. Other sources of information on planetary composition and history are the Moon and comets. The Moon shows a history of accumulation and segregation, as well as impacts by meteoritic materials and volcanism, but it contains no evidence of an extensive atmosphere of any sort. Comets are essentially all "atmosphere" in that they are composed of frozen water and organic compounds as well as silicate materials. If impacted on a planet a comet will provide an "instant" atmosphere even if only transiently.

THE STRUCTURE OF THE EARTH

The Earth is a radially zoned planet. The surficial zone is visible to us as land and sea. The distribution of elevations of rock below and above sea level indicates that the average continental height is about 800 meters above sea level and the average depth of the oceans is about 4,000 meters.

The continents all have large ancient areas of granitic and metamorphic rocks around which, in irregular fashion, are distributed younger granitic and metamorphic rocks as well as sedimentary rocks (limestone, shale, sandstone) and volcanic rocks. The oceans, on the other hand, covering 71 percent of the Earth's surface are underlain by volcanic rocks called basalt, similar to those found on Hawaii or Iceland, with a superficial veneer of sediments.

We call the outermost layer of the Earth the *crust.* To determine its thickness and what lies in the deeper zones of the Earth, we must have a way of seeing through about 6,000 kilometers of planetary material. This is accomplished through the use of earthquake waves penetrating the Earth's interior, and the instrument used is called a *seismograph.*

When an earthquake occurs, the energy of a small nuclear blast (about 10^{25} ergs) is released to the Earth. (The actual point where the energy is released is called the *focus* and the point on the Earth's surface directly above the focus is called the *epicenter.*) This energy is transmitted through the Earth as waves and the laws governing waves determine the paths they take. The waves can be *reflected, diffracted,* or *refracted,* and all along the way their energy lost to the medium. Of interest in exploring the layering in the Earth are the phenomena of *reflection* and *refraction.*

Reflection is a common experience with waves: light rays (a line perpendicular to the wave front is called a *ray*) reflecting from a hard nonabsorbing surface called a mirror; sound waves reflecting from the hard nonabsorbing walls of a canyon resulting in an echo. The primary use of this sort of information in the Earth is for the study of the superficial sediment layers in the ocean, as we shall see later.

Refraction is the phenomenon observed when waves cross a boundary between two media in which the wave has different velocities. This change in velocity causes the wave front (and thus the ray) to bend. The ray approaching the boundary is called the incident ray and the ray after passing through the interface between the two media is called the refracted ray. The angle between the incident ray and the normal is called the angle of incidence (i) and the angle made by the refracted ray and the normal is called the angle of refraction (r). (The *normal* is a line perpendicular to the boundary.) It can be shown that:

$$\frac{\sin i}{\sin r} = \frac{V_i}{V_r}$$

where V_i and V_r are, respectively, the velocity of the wave in the original medium and the velocity in the medium being entered by the ray. This is called Snell's Law. Clearly if V_i is greater than V_r then $\sin i$ will be greater than $\sin r$ and i will be greater than r.

The refraction of the energy waves from an earthquake passing through the Earth determines how the earthquake will be recorded on seismographs deployed around the Earth's surface. On the basis of the records of thousands

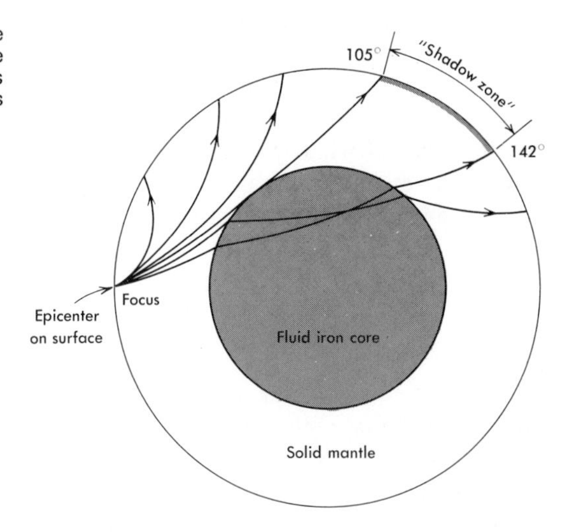

FIG. 1-1 The path of earthquake rays and the inferred interior structure of the earth. The outer core behaves like a liquid and the inner core behaves like a solid.

of seismographs all over the world, a self-consistent picture of the ray paths through the Earth emerges (see Fig. 1-1). The Earth is thus divided into three major zones—proceeding in from the surface: the *crust*, the *mantle*, and the *core*.

The boundary between the crust and the mantle is called the Mohorovičić discontinuity because a major change in velocity occurs at the boundary. This change in velocity is interpreted as due to a radical change in rock type between the crust and the underlying mantle. The velocity increase with depth in the mantle is gradual without marked discontinuities. It is due to the increasing pressure with depth, which increases both the density and the elastic constants of the medium. The density and elastic constants together determine velocities.

Table 1-2 Major Subdivisions of the Earth

Unit	Depth (kilometers)	Mass ($\times 10^{24}$ grams)	Volume ($\times 10^{10}$ cubic kilometers)	Density (grams per cubic centimeter)	Percent Mass of Earth
CRUST					
CONTINENTAL	0–40	16			
SEA WATER	0–4	1.4			
OCEANIC	4–11	7.0			
AVERAGE		25	0.9	2.8	0.4
MANTLE	25–2,900	4,075	89.4	3.5 increasing with depth to 6.5	68.2
CORE	2,900–6,370	1,880	18	9.5 increasing with depth to 13.5	31.4
EARTH		5,980	108.3	5.517	100.0

A marked decrease in velocity of the medium occurs at the mantle-core boundary. On the basis of the average density of the Earth and of mantle rocks, it is evident that the core must have a much higher density than the mantle. The material that best fits this requirement is an iron-nickel metal. This is reasonable on the basis of what we know about the occurrence of metallic iron-nickel in meteorites and the generally high cosmic abundances of these metals.

Table 1-2 gives the best current description of the structure of the Earth.

THE HYDROSPHERE

The Earth is like a four-ring circus with acts peculiar to each ring going on simultaneously. Solid rocks and sediments are made, destroyed, and moved around in the *lithosphere**; water permeates the rocks, fills basins, evaporates, condenses, and flows in channels—this realm is called the *hydrosphere*; gases surrounding our planet compose the *atmosphere*; and all of life in whatever environment it finds itself is called the *biosphere.* Unlike the rings of a circus, the four Earth "spheres" are not isolated from each other. One influences the other and at times the boundaries are not easily discerned.

We are concerned with the processes and events that involve the major repository of the hydrosphere—the oceans. We cannot study the ocean even as merely a reservoir of water without understanding its relation to the rest of the hydrosphere.

The Properties of Water

Pure water, that is, water free of any dissolved substance, is a remarkable material from any view. It is in a sense the perfect solvent, since everything has some solubility, no matter how small, in it. It is a transparent fluid even with a burden of dissolved material, permitting light penetration to some depth (Fig. 1-2). This is of importance to life in the sea. Water has the highest heat capacity of any common substance, which means that large additions or subtractions of heat do not change the temperature very much compared to an equal mass of iron or rock. This heat capacity acts as a regulator of atmospheric temperatures in maritime lands. Water also has a high latent heat of vaporization and melting. This means that large amounts of energy or heat are required to vaporize liquid water or melt solid ice—much more per unit mass than anything else on Earth. Thus the movement of water and conversion from one phase to another involves an efficient method of energy transport. A not altogether welcome example of this is the energy of hurricanes, which derives from the condensation of liquid water in the atmosphere. The high water vapor content

---

* The term *lithosphere* is used here in a broad chemical sense. It is defined and used in Chapter 8 in the tectonic sense. The two usages should not be confused.

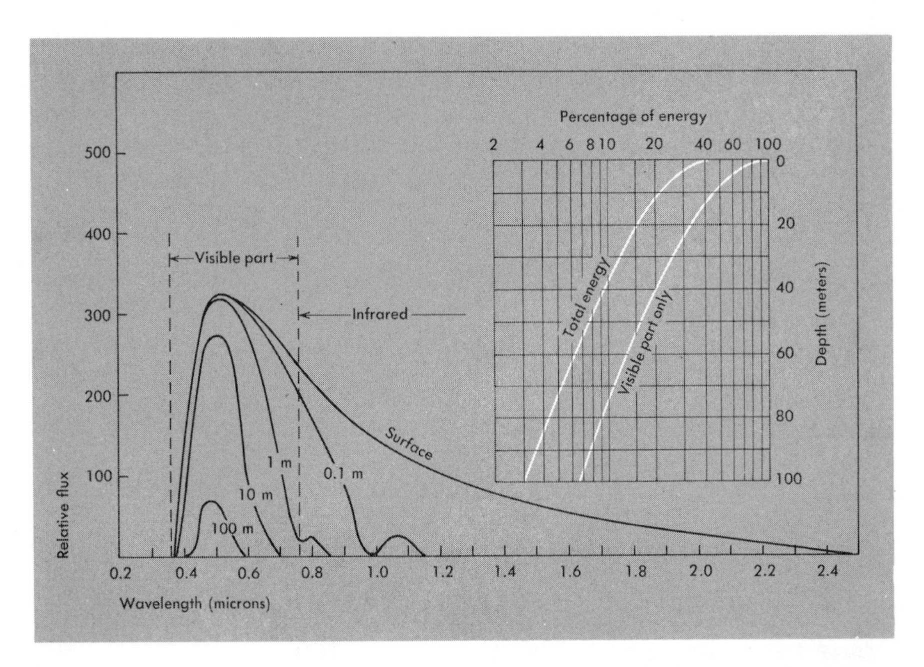

FIG. 1-2 Representation of relative fluxes of light as a function of wavelength, after passage through increasing depths of pure water. Note that the total incident light energy at 100 meters is less than 3 percent of the energy at the surface and that the light is primarily toward the blue end of the spectrum. (From Sverdrup, Johnson, and Fleming, 1942.)

is supplied by evaporation from the sea in tropical areas. The energy of the Sun required to do this is released once the water condenses. If this is done in confined areas the large energy release has the massive force that we call a hurricane.

A peculiar property of pure water is its maximum density as a liquid at 4°C (Fig. 1-3). In a standing body of water such as a lake, this will result in sinking so that the whole lake will overturn as cooling from the atmosphere continues. Eventually, when the whole lake is at 4°C, the surface waters will cool further. As the density becomes progressively lower between 4°C and the freezing point at 0°C, the cooling lower density water will remain floating on

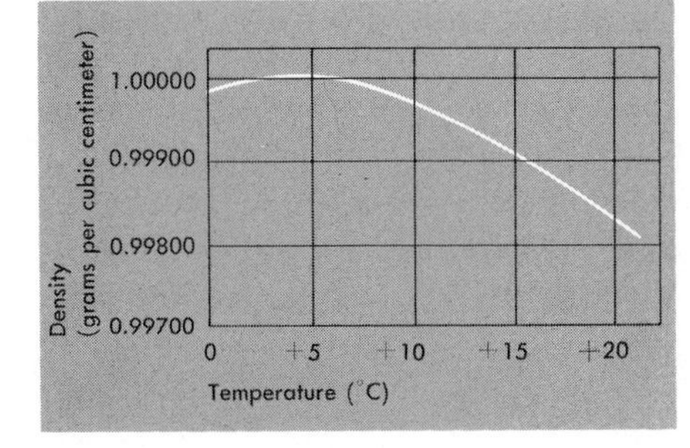

FIG. 1-3 The variation of density with temperature for pure water at one atmosphere pressure. The maximum density occurs at 4°C and freezing occurs at 0°C.

the higher density 4°C water and eventually freeze. Because of the high latent heat of freezing (that is, because a lot of heat has to be extracted from liquid water to make ice), the lake will rarely freeze right down to the bottom. The fact that ice, the solid state of water, is less dense than liquid water and floats permits the survival of fishes in winter. At sea it also results in iceberg hazards to shipping.

The Hydrologic Cycle

The ancient author of the Book of Ecclesiastes well understood the hydrologic cycle when he said: "All the rivers run into the sea; yet the sea is not full; unto the place from whence the rivers come, thither they return again."

The major reservoirs of water on the Earth are shown in Table 1-3.

Table 1-3 Relative Distribution of Water in the Earth's Surface Other Than in the Oceans and Excluding Rocks

Reservoir	Percent	Volume (cubic kilometers)
POLAR ICE AND GLACIERS	75	29×10^6
GROUND WATER AT DEPTHS LESS THAN 750 METERS	11	4.2×10^6
GROUND WATER AT DEPTHS GREATER THAN 750 METERS, BUT LESS THAN 4,000 METERS	13.6	5.3×10^6
LAKES	0.3	120×10^3
RIVERS	0.03	12×10^3
SOIL MOISTURE	0.06	24×10^3
ATMOSPHERE	0.035	13×10^3
TOTAL NONOCEANIC WATER	100	39×10^6
VOLUME OF OCEANS		$1{,}350 \times 10^6$

Without question the most important is the oceanic reservoir. Of the remaining reservoirs making up less than 10 percent of the oceans, the water tied up in the rocks of the crust and water in the form of ice in Antarctica and Greenland primarily are of about equal importance. All the other reservoirs are minute in comparison, although for the majority of us they represent our most tangible experience with the realm of water.

The reason the other reservoirs are so important in human experience is that although they are small in volume the water is perpetually on the move here. Rain and other forms of precipitation transport the water vapor in the atmosphere—derived ultimately from the ocean—to land, where it flows down rivers or sinks into the ground to form ground water before making its final journey to the sea. Some of the water on the continents will also evaporate back into the atmosphere before it has a chance to flow to the sea.

Table 1-4 Water Balance on the Earth's Surface

Process	Grams per Year × 10^{20}
EVAPORATION FROM OCEAN	3.83
PRECIPITATION ON OCEAN	3.47
EVAPORATION FROM LAND	0.63
PRECIPITATION ON LAND	0.99
RUNOFF FROM LAND TO SEA	0.36

Table 1-4 is an estimate of the water fluxes in the hydrologic cycle. If we thought that the streams derived their water from some mysterious source and were filling up the oceans with water, we could calculate the time necessary to fill the oceans to the present level. From Table 1-4 we see that the rate of supply of water by streams is about 36×10^{18} grams per year (since the density of water is about 1, this is equivalent to 36×10^{18} cubic centimeters per year or 36×10^{3} cubic kilometers per year). The present-day volume of the oceans is about $1{,}350 \times 10^{6}$ cubic kilometers; thus:

$$\text{length of time for ``filling''} = \frac{\text{volume of the oceans}}{\text{rate of supply by streams}}$$

$$= \frac{1{,}350 \times 10^{6}\ \text{km}^{3}}{36 \times 10^{3}\ \text{km}^{3}\ \text{per yr}} = 37{,}500\ \text{yr}$$

With our uncertainties we can round this off to 4×10^{4} years. Since the oceans are cycling with the atmosphere and surface runoff, this cannot be a significant estimate of filling-up time. Rather, it is a measure of the cycling rate. The value of 4×10^{4} years is the *mean residence time* of water in the oceans before it experiences evaporation, precipitation on land, and return via streams to the ocean. Some water molecules may be cycled more frequently than others, but on the average that is the residence time of water molecules in the ocean relative to the continental resupply pathway.

HOW DID THE EARTH GET ITS OCEAN AND ATMOSPHERE?

If Earth is uniquely the "water" planet, how did it get that way? Any question that we ask about the source of the water in the oceans might as well be asked about the chloride in sea water, carbon dioxide and nitrogen in the air, and organic compounds in animals and plants. For none of these compounds in the quantities observed can be derived from the weathering of granite and basalt, which make up most of the present-day crust of the Earth.

Water vapor, hydrogen chloride gas, carbon dioxide, carbon monoxide or methane, and nitrogen or ammonia gas would be excluded in the growing Earth as gases. Only the nonvolatile compounds of chlorine, carbon, hydrogen, and nitrogen would accumulate during the formation of the Earth. Water in clay-type minerals and other hydrated silicate minerals; carbon and nitrogen in organic compounds with low boiling points; and chlorine as ions substituting for the oxygen atoms in silicates or occurring as chloride minerals—these are the forms that would have permitted accumulation of these "volatile" elements. All of these compounds have been found in certain meteorites, known as the *carbonaceous chondrites*. The incorporation of carbonaceous chondrite-type material at some stage in the growth of the planet, with subsequent heating and degassing, would provide the atmospheres for the terrestrial planets. The question remains as to how much and when was carbonaceous chondrite-type material added to the planet.

For a number of reasons, based on our knowledge of the composition of rocks derived from the Earth's mantle and the information from the Moon and meteorites, it appears that we can treat the accumulation of the terrestrial planets as a nonhomogeneous process. As the Sun condensed and the planetary disk formed around this rotating body, the temperature in the region of the inner planets increased to about 2,000°K and much of the material was in vapor state at one-thousandth of the pressure of our atmosphere. As the nebula began to cool the high temperature minerals condensed and accumulated, making up the inner parts of the terrestrial planets. Melting and fractionation occurred within the planet as the result of heating primarily from gravitational energy. This all must have occurred in less than a million years. Up to that point no volatiles in any form would have accumulated. These were "dry" planetary nuclei. The planets were then veneered, more or less depending on the planet, with carbonaceous chondrite-type material, which carried the potential volatiles to the surface of the planet. Subsequent melting and recrystallization in the outer parts of the planet released the volatiles to the surface and if the planetary body was large enough they would be retained.

Venus and Earth appear to have formed with the same amount of volatile-containing veneer, whereas Mercury and the Moon, apparently have not had any volatile-rich veneer added at any time. On the Moon this is clearly seen in the virtual absence of volatile metals such as lead and mercury, which are plentiful on the Earth. Mars is an intermediate case. It has an atmosphere that indicates the accumulation of a veneer, but possibly because of the planet's size some of it may have escaped.

The oldest rocks on Earth have been dated at about 3.7×10^9 years whereas the age of the Earth is about 4.6×10^9 years. Thus, the first 900 million years of Earth history as recorded in rocks have been obliterated by subsequent events. In the absence of any record of the early history of the Earth's surface, we can only project the consequences of the model we chose for the origin of the planet.

If the Earth's volatile-bearing veneer accumulated in a relatively short time early in its history, as appears to be the case from geochemical arguments, then a dynamic upper mantle and crust system would have developed very soon after the accumulation process, due to the extensive energy release and heating of the exterior of the planet. At this time most of the volatiles and the compounds soluble in the fluids being transported would be transported to the surface of the Earth. Thus we would have an ocean and an atmosphere soon after the planet formed.

Under these conditions the volume and saltiness of the oceans would have been established early. The initial atmosphere, however, would be very different from that of today. It would contain primarily carbon dioxide, nitrogen, and water vapor. With time the dissociation of water by ultraviolet light would build up a small reservoir of oxygen, which would be reactive with rocks at the surface. The carbon dioxide pressure would diminish, possibly to as low as a thousandth of the present-day carbon dioxide pressure, by reaction with rocks. Only with the advent—perhaps within the first 1,000 million years of Earth history—of photosynthesis were the oxygen and carbon dioxide contents of the atmosphere built up to and maintained at approximately the present-day levels. By this model the oceans were essentially established at the beginning of the Earth, and subsequent degassing of the planet has been of minor importance.

The idea of a continuously degassing Earth has also been proposed. We know that volcanoes, geysers, and fumaroles release huge quantities of gases and water into the atmosphere. Most of these, however, are recycled materials (water, carbon dioxide, hydrogen sulfide, and hydrogen chloride), which are easily derived from the alteration of sediments and percolating waters from the surface of the planet.

There is, however, clear evidence of degassing from newer data on the distribution of rare gases in sea water and in glassy volcanic rocks found in the ocean deep. These findings show a small continuing flux of helium, neon, argon, krypton, and xenon from the mantle to the Earth's surface. The fluxes of reactive gases like water and carbon dioxide, however, are small when calculated on the basis of these data.

topography and structure of ocean basins

The ocean basins are not merely inert receptacles for water and sediment. The topography and structure of the ocean bottom are highly variable from place to place and reflect processes in the Earth's interior. These features also vary in permanence so that the ocean bottom of today is not like the ocean bottom of 50 million years ago. In this chapter we will survey what has been learned of the topography and structure of ocean basins, mainly as the result of intensive studies during the last 30 years.

OCEAN-BOTTOM TOPOGRAPHY

For determining the major topographic features of the ocean bottom, some sort of remote sensing device is required. Shallow water depths were determined in the past by means of a *sounding line* made of hemp with a lead weight attached (hence the phrase "plumbing the depths" from the Latin *plumbum* for lead). In the latter part of the nineteenth century, hemp gave way to metal wire, and until about 1920 soundings were obtained in all parts of the oceans by this method. In shallow waters, soundings could be made relatively rapidly, but in the deep ocean, where the average depth is 4,000 meters, it would take several hours to lower and raise the sounding line. When research vessels began to dredge the deep ocean depths in the latter part of the nineteenth century the length of cable required to lower the dredge to the ocean bottom gave additional depth information. Similarly, deep-water sampling programs could be used for determining depth at the location of sampling. It is evident that gathering data on depth distribution in the oceans was a long, tedious, and spotty endeavor when these methods were employed. Nevertheless, by 1920 enough soundings had been obtained to indicate the elementary fact that the deep ocean bottom is not a featureless plain and to permit the cataloging of the distribution of depths at sea.

After 1920, although the sounding-line method continued to be used in oceanographic programs, *echo sounders* began to be developed. This technique depends on generating a sound signal that is transmitted through the water, reflected off the bottom as an echo, and received aboard the ship generating the signal. Since the velocity of sound is a function of the temperature, salinity, and pressure of the water, these properties must be known or approximated in order to translate the transit time of the signal into depth. During World War II these techniques were immensely improved, and since then all soundings have been made with continuously operating, precision depth-recorders that give bottom profiles of the type seen in Fig. 2-1.

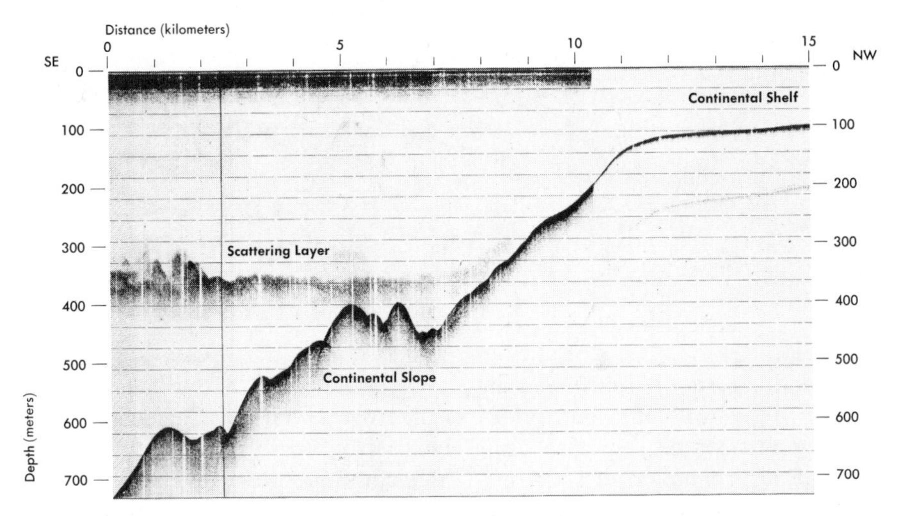

FIG. 2-1 Precision echo sounding across the continental margin off Virginia. The record shows the transition from the nearly horizontal smooth surface of the outer continental shelf to the inclined irregular surface of the upper continental slope. Reflections from a deep scattering layer within the ocean are visible between 300 and 400 meters below sea level. The scattering layer is believed to be due to certain organisms. Vertical exaggeration is about 12:1 so that the true inclination of the upper continental slope is about 6° from horizontal. (Courtesy Peter A. Rona.)

Methods of navigation have also improved considerably since the war, thus permitting the accurate location of ocean-bottom features. Standard celestial and solar navigation is still a major means of getting accurate fixes of location. But since this method requires a precise measurement of the distance of a heavenly body above the horizon, overcast skies are a hindrance. In such cases a ship is forced to proceed by dead reckoning based on the position calculated from the last astronomic fix. Accuracy is diminished because of ocean currents and drift due to winds. Now, however, at least within transmitting distances

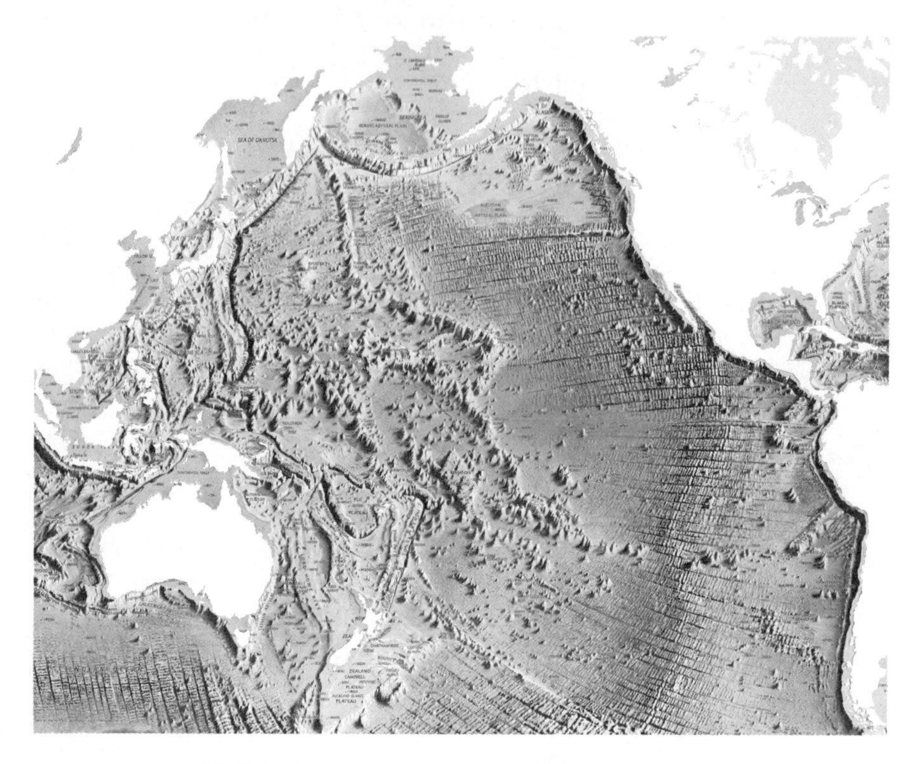

FIG. 2-2 Physiographic diagram of the Pacific Ocean floor.

from land, a form of electronic triangulation called LORAN can be used in all weather, permitting an accuracy of less than one kilometer. Currently, too, a growing number of ships are using signals from artificial satellites for the accurate determination of location. This method is particularly important in remote areas of heavy overcast skies, such as the Antarctic, because a purely electronic signal from a precisely known source is used in establishing location.

The results of the very large number of soundings made of the ocean bottom indicate that there are three major topographic features common to all oceans: *the continental margins*, *the ocean-basin floors*, and the major *oceanic ridge systems*. In addition to these major features are additional ridges and rises that occur throughout the ocean basins. The general physiographic map shown in Fig. 2-2 provides an idea of how the topography of the ocean basins varies from place to place.

The Continental Margin

The continental margins around the Atlantic Ocean Basin are commonly composed of the sequence shelf-slope-rise proceeding seaward (as in Figs. 2-1 and 2-2). The *continental shelf* is the submerged continuation of the topography and geology visible on the adjacent land, modified in part by marine erosion or sediment deposition. The *shelf break*, wherever it can be seen unambiguously, marks the seaward extent of the continental shelf and occurs at depths of between 10 to 500 meters, averaging 200 meters. The relief on the shelf is generally less than 20 meters. The *continental slope*, the next seaward feature, is demarked by the shelf break on the landward side; here the gradient changes from 1:1,000, typical of the continental shelf, to steeper than 1:40. The continental slope may include a series of escarpments. The *continental rise*, at the foot of the continental slope, has gradients ranging from 1:1,000 to 1:700. In many areas, it is a depositional feature. *Canyons*, such as the Hudson Canyon (Fig. 2-3), cut across the continental rise and act as channels for the seaward transport of sediment. Some canyons have continuations on the continental slopes and shelves, but many do not.

FIG. 2-3 Preliminary chart of Hudson Submarine Canyon, based on nonprecision soundings taken 1949–1950. (From Heezen, Tharp, and Ewing, 1959, by permission of The Geological Society of America.)

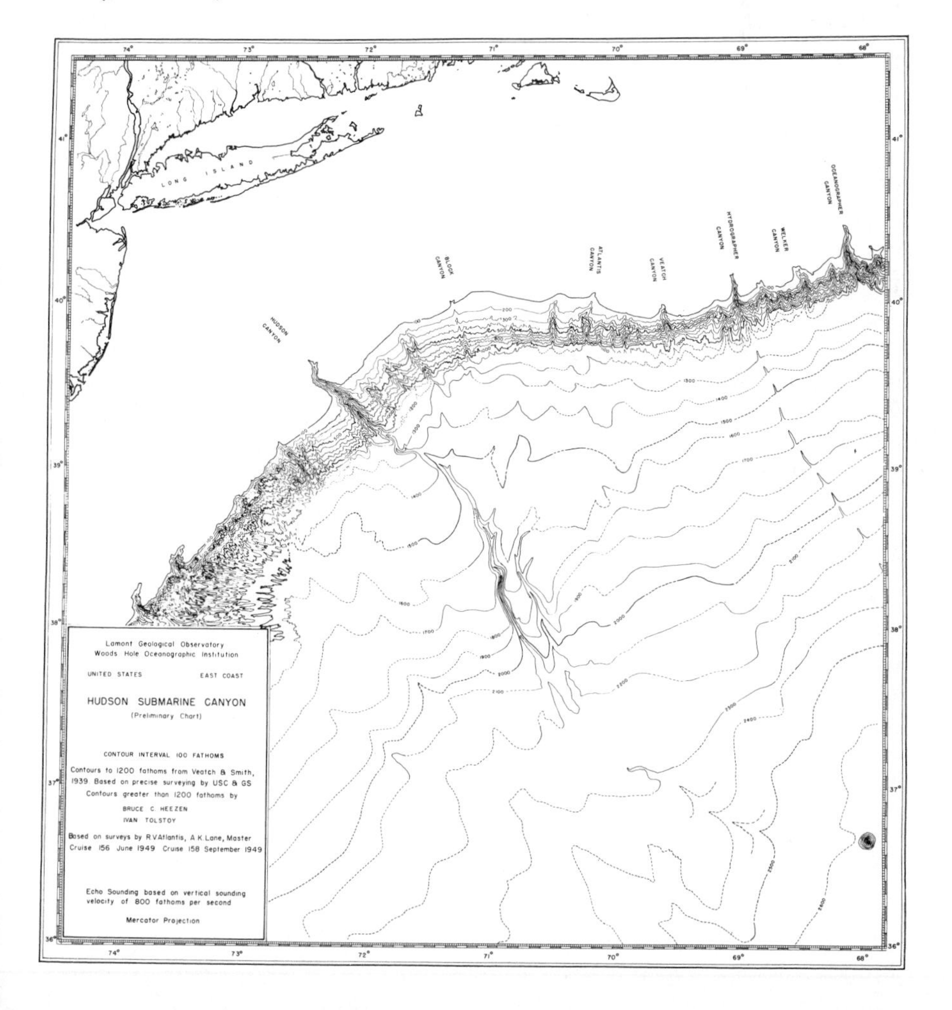

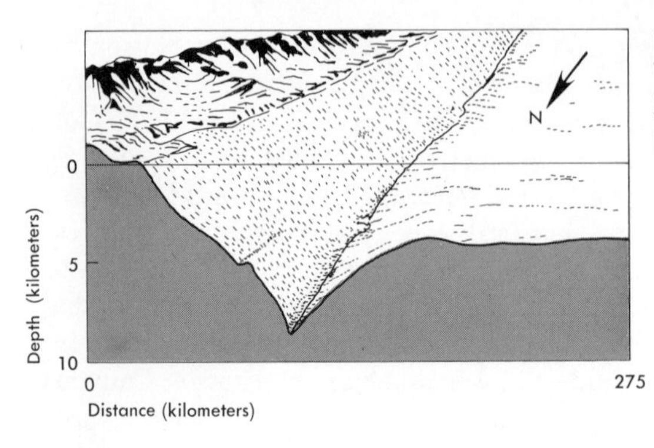

FIG. 2-4 A block diagram of the trench off Antofagasta, Chile. The mountains to the east are the Andes and the trench is 8 kilometers deep. (After R. Fisher and R. Raitt, 1962.)

In many parts of the world, the continental margin beyond the shelf consists of a deep trench (Fig. 2-4). These trenches, especially prominent around the Pacific Ocean Basin, are the deepest parts of the oceans, often exceeding eight kilometers in depth. In other parts of the Pacific margin the continental margin as a whole is a continuation of the structures found on the adjacent land.

The Ocean-Basin Floor

The ocean-basin floor includes everything seaward from the continental margin except for the major oceanic ridge systems. The floor makes up one-third of the Atlantic and Indian ocean basins, and three-quarters of the Pacific Ocean Basin. *Abyssal plains* adjacent to the continental rises—widespread in the Atlantic and sparse in the Pacific—are very smooth, with gradients between 1:1,000 and 1:10,000. All abyssal plains are connected by canyons or other channels to landward sources of sediments, which are transported as dense slurries and deposited on the plains. The Argentine Basin has a rugged topography shaped by currents from a thick pile of sediments. Most of the Pacific Ocean floor is occupied by abyssal hills forming a hummocky topography. These hills are probably caused by volcanic intrusions and extrusions. *Seamounts*, which do not penetrate the surface of the ocean, and *oceanic islands* such as the Hawaiian Islands are also of volcanic origin. Figure 2-2 shows the widespread occurrence of these volcanic features. In the Pacific Ocean the volcanic islands and seamounts commonly form clusters connected deep under the surface of the ocean by the accumulations from volcanic activity and the erosion of volcanoes. Topographically they look like featureless aprons around the volcanic peaks. Flat-topped seamounts, or *guyots*, occur throughout the oceans and indicate that the tops were once above sea level. The "flattening" is due to erosion at sea level. Some of these guyots occur at depths of 1,600 meters.

The Major Oceanic Ridge Systems

The major oceanic ridge systems form a series of connected, topographically high areas present in all the oceans (Fig. 2-2). Ridges are between 1,000 and

4,000 kilometers wide with a relief of two to four kilometers above the ocean floor, at points protruding from the sea surface as islands. The term "midoceanic ridge" has sometimes been used for the system, after the most prominent example, the Mid-Atlantic Ridge. The topography is representative of a composite of volcanic and rupture features called faults. At the center of the Mid-Atlantic Ridge, for example, is a discontinuous "rift valley" characterized by heavy earthquake activity and higher-than-average heat flow. A series of transverse trenches that offset the axis of the ridge is also prominent. Other oceanic ridges, such as the East Pacific Rise, have many of the elements of the Mid-Atlantic Ridge but not necessarily all. Nevertheless, these differences aside, the ridges appear to be continuous around the Earth, except for offsetting by transverse breaks.

OCEAN-BOTTOM STRUCTURE

Having obtained a topographic picture of the ocean bottom, we can ask the next logical question: What are the physical properties and dimensions of the various types of rock and sediment that act as the "liners" of the ocean basin? To answer this question we must turn to the methods of geophysical exploration.

Seismic Refraction

The same principles that have been used to study the interior structure of the Earth by means of earthquakes, as described in Chapter 1, can also be applied to the study of the structure of ocean basins. Earthquakes provide a large source of energy that is propagated as sound waves for great distances through the Earth's interior. The times of arrival of the different types of waves at various locations on the Earth, as measured by seismographs, can be used to determine the internal structure of the Earth. On a much smaller scale, man-made explosions can be used to generate sound waves that provide detailed information about the structure of the surface layers of the Earth, and it is this method that has been used successfully to determine the structure of ocean basins.

The technique of seismic refraction at sea makes use of sound waves supplied by an explosive set off from a ship. The sound waves penetrate the ocean bottom. The sediments and rocks of the ocean basins form a layered sequence in which each successively deeper layer has a higher velocity of propagation of sound waves than the one above. With such a configuration, the method of seismic refraction gives reasonable interpretations of the thickness and composition of each of the main layers.

Although sound may be transmitted in several different ways through the Earth, the ocean, being liquid, can only support sound waves that are transmitted by a process of squeezing and release (compression and rarefaction). These waves are called compressional waves and their velocities in various media are designated as V_p. (The subscript p refers to an early designation for the primary waves, which are the first waves, passing through the Earth's interior, that are recorded by a seismograph when an earthquake occurs.)

Since the various layers of the ocean bottom have different compressional-wave velocities, a particular sound ray from an explosion will be refracted at the interface of two layers into the plane of the layer it has just entered. As it moves away from the explosion source, it is continuously returned upward to the ocean surface until all its energy is dissipated. At any determined point away from the explosion, the first large signal on a recording instrument will be due to the sound traveling through the fastest path. The second pulse will be from sound traveling through the next fastest path and so on. In order to get enough data to define clearly the various major strata, records of arrival times are obtained at several locations away from the explosion. The waves are detected either aboard another ship whose distance from the moving "shooting" ship is changed between explosions or by a string of sound detectors, called hydrophones, behind the ship (Fig. 2-5).

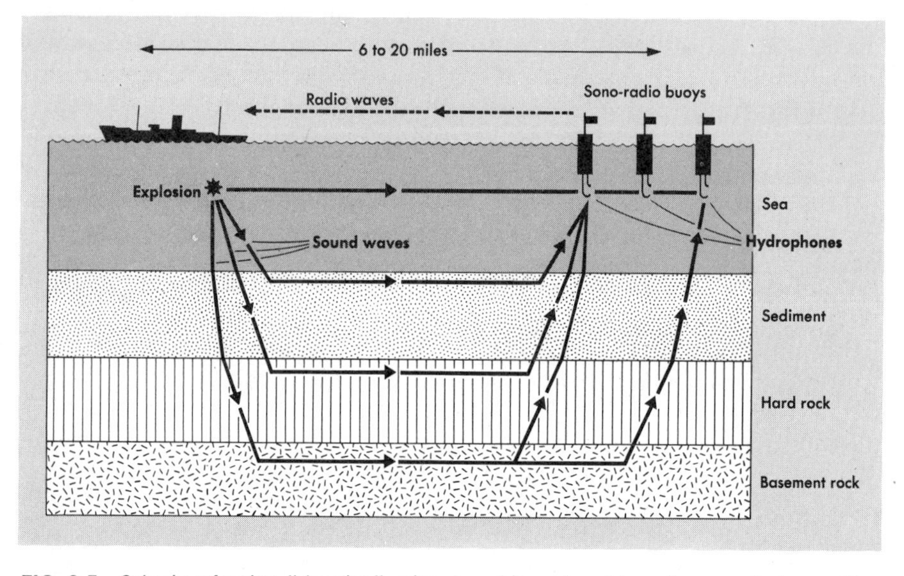

FIG. 2-5 Seismic refraction "shooting" using one ship and a string of sonoradio buoys. The relative dimensions of the ship, buoys, and the various layers are exaggerated. Seismic refraction studies are also made with two ships, one shooting and the other receiving. The sound waves generated by the explosive are refracted at the boundaries of the different layers. For the sequence of materials normally encountered in the deep-sea layers, the velocity of sound increases progressively downward from layer to layer, causing a return of the refracted signals at each interface. (After Hill, 1957.)

The general structure of the deep ocean basins, as determined by using seismic refraction, is represented by the cross section across the Atlantic shown in Fig. 2-6. Table 2-1 lists the properties of a generalized layering in the oceanic crust.

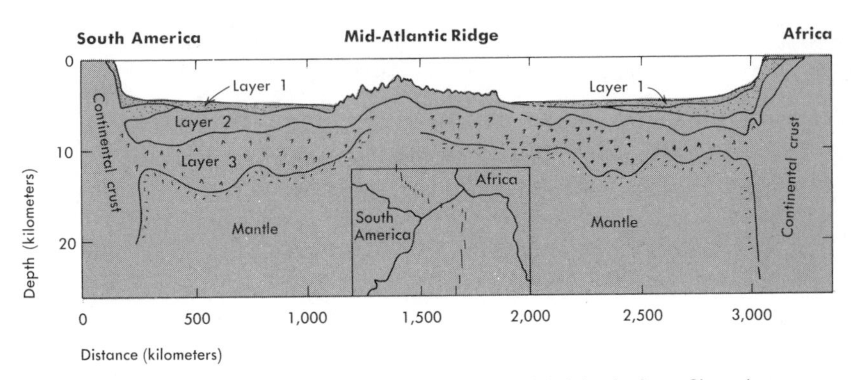

FIG. 2-6 Seismic refraction section across the equatorial Atlantic from Sierra Leone to Brazil. (After Leyden, et al., 1967.) The properties of the layers are described in Table 2-1.

Table 2-1 Thickness and Properties of the Layers of the Ocean Floor

Layer	Approximate Thickness (kilometers)	Velocity of Compressional Waves, V_p (kilometers per second)	Likely Material	Approximate Density (grams per cubic centimeter)
SEA WATER	4.5	1.5	Sea water	1.0
LAYER 1	0.45	2.0	Unconsolidated sediments	2.3
LAYER 2—BASEMENT LAYER	1.75	4.0 to 6.0	Consolidated Sediments or volcanic rocks	2.7
LAYER 3—OCEANIC LAYER	4.7	6.7	Basalt (in part altered)	3.0
MOHOROVIČIĆ DISCONTINUITY				
LAYER 4—MANTLE	—	8.1	Ultrabasic rocks	3.4

Gravity Measurements

To a first approximation, the acceleration of gravity is constant over the Earth's surface. That is, a rock falling from a thousand-meter height any place in the world should have accelerated to the same velocity at the moment of impact with the ground. Similarly, a pendulum would have the same period all over the Earth if our approximation is valid, since the period is directly related to the acceleration of gravity. However, the period of a pendulum that is moved around the Earth, is greater at the equator than at higher latitudes.

This fact was discovered when properly working pendulum clocks made in Europe were brought to the Caribbean and equatorial South America in the eighteenth and nineteenth centuries; they invariably ran slower than they did in the European homeland. This indicates that the acceleration of gravity decreases with decreasing latitude. Newton's law of universal gravitation states that the force of attraction (F) between two bodies is directly proportional to the product of the masses of the two bodies (M_1, M_2) and inversely proportional to the square of the distance (R) between their centers of mass:

$$F = G\frac{M_1M_2}{R^2}$$

where G is the proportionality constant.

On a rotating body like the Earth the force due to gravitational attraction is in part offset by the centrifugal force due to rotation. The centrifugal force of the rotating Earth is greater at the equator than at the poles, thus causing it to bulge at the equator and flatten at the poles. Due to this bulging the distance from the center to the surface of the Earth at the equator is greater than at the poles or any point between. This means that the gravitational acceleration will be diminished at the equator relative to the poles. This effect, however, is responsible for only 35 percent of the difference in gravitational acceleration between the poles and the equator. The other 65 percent is due to the fact that a free body on the surface of the Earth is subject to the centrifugal force of the rotating planet. This force is in the opposite direction to the force due to gravity, thus subtracting from it. The strength of this counterforce is greatest at the equator; thus the acceleration of a falling body is least at the equator.

Using this information we can calculate the expected acceleration due to gravity at any latitude. The presence of large excesses of mass such as mountains, or deficiencies such as deep basins, at the Earth's surface will alter the acceleration of gravity locally and the attractive force of a nearby object can be sensed by highly accurate instruments.

The modern-day geophysicist uses a very sensitive pendulum or a spring called a gravimeter to measure the variation in the acceleration of gravity over the Earth's surface. It records the effects of local excess or deficiency of mass in modifying the acceleration of gravity that is predicted for a given latitude from the Earth's figure. Until the 1960's the only stable platform at sea on which a pendulum or gravimeter could be operated and give significant results was aboard submarines below the action of the ocean waves. But now, with proper instrument platforms, gravimeters can easily be used on surface ships.

Gravity measurements at sea show differences due to local variations in mass distribution resulting from the varying thicknesses of the layers of the oceanic crust and upper mantle. The gravity results are best used in conjunction with seismic data and with the measured or inferred density of the various layers that have been determined seismically. Along the continental margins, for example, the location of the Mohorovičić discontinuity is not clearly delineated by seismic data. Although the discovery of this major worldwide discontinuity was made on the basis of seismic refraction studies, we know that it is also a discontinuity between material of different densities as seen in Table 2-1. The depth of the Mohorovičić discontinuity thus may be inferred from gravity data, by assigning densities to the major rock or sediment layers and determining the thickness of each layer that would be compatible with the local gravity measurements.

Crustal structures of the oceanic ridge systems are also inferred from both seismic and gravity data. For instance, the seismic data indicate that, in addition to the layers in the "normal" section described in Table 2-1, underneath the axis of the ridge a layer with a velocity of about 7.3 kilometers per second is encountered in the mantle. Normal mantle velocity under both continents and oceans is about 8.1 kilometers per second. The low velocity of the mantle layer under the ridge is due to less dense material (possibly molten rock formed from ultrabasic rocks of the mantle). The gravity data give dimensions of the various velocity layers when an appropriate density, based on the inferred rock type, is assigned to each layer. These results on a section across a typical part of the major ocean ridge system are shown in Fig. 2-7.

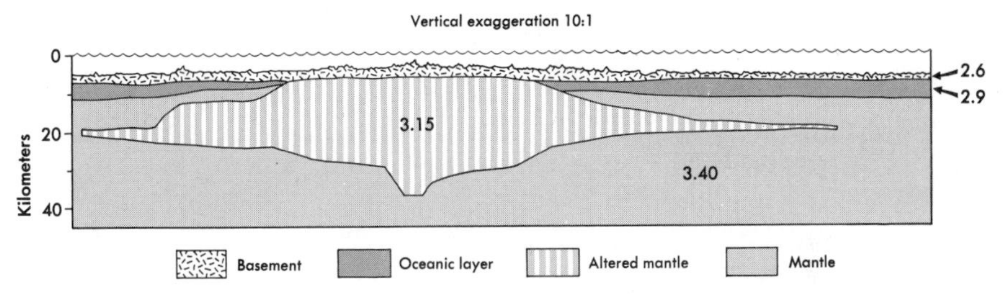

FIG. 2-7 A possible crustal model across the north Mid-Atlantic Ridge that satisfies both the gravity data and the seismic data. The inferred densities used in the model are given on the figure for each unit. The 3.15 density rock underlying the ridge area is the "anomalous" mantle, referred to in the text, having a seismic primary wave velocity of 7.3 kilometers per second compared to the normal mantle velocity of 8.1 kilometers per second. The sediment and sedimentary rock layers are too thin along the ridge to appear in the diagram. (After Talwani, LePichon, and Ewing, 1965.)

Seismic Reflection—The Structures and Thicknesses of the Sediment Layers

Seismic refraction data, although useful for delineating the major units of the deep-sea floor, give very little information on the sedimentary layer ("Layer 1") itself. Seismic reflection, however, is more sensitive to the presence of differences in sedimentary layering that result in distinct reflection horizons for sound waves. The techniques are similar to the method of seismic refraction except that generally smaller energy sources are required and sound detectors must be close to the ship emitting the signal in order to record the waves bouncing from acoustic reflectors in the sediments.

It had been observed before 1961 that, in both seismic refraction work and continuous echo sounding, reflection horizons in sediments could be detected. By this method a continuous layer of volcanic ash was discovered in the East Pacific, for instance. Since 1961 the technique of using *seismic reflection* at sea has been perfected for investigating sedimentary structure and thickness. The energy pulses are supplied by explosive charges (for deep penetration into the sediment), by electric spark "pingers" (similar to the equipment used in obtaining echo soundings but at different frequencies so as to diminish attenuation by sound absorption by the sediments), and more recently by compressed air. Equipped with these devices, oceanographers have explored the structure and thickness of sediments of the major oceanic areas.

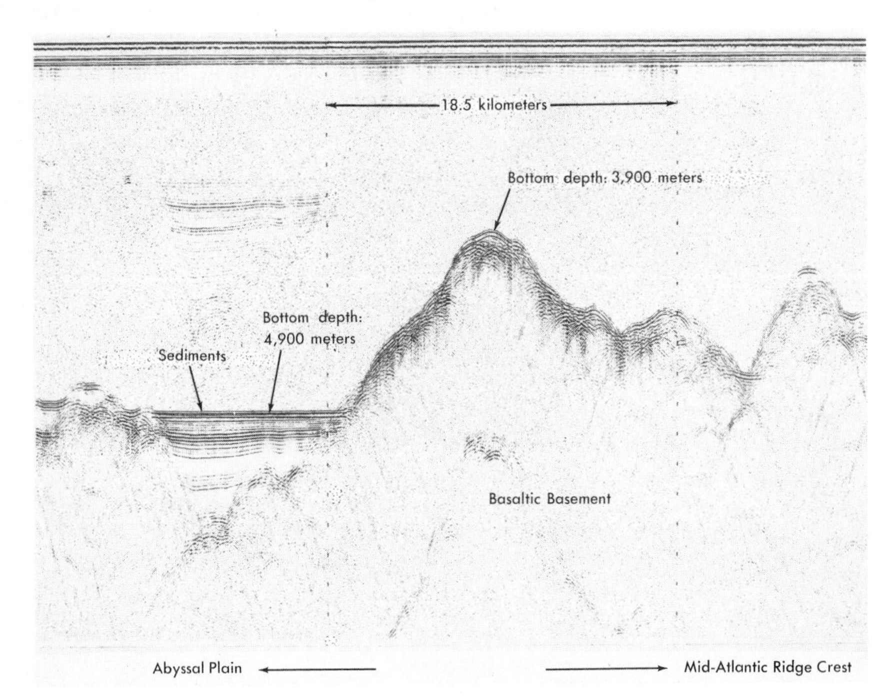

FIG. 2-8 Seismic reflection profile of a portion of the Mid-Atlantic Ridge western flank or rise east of Trinidad (9°30′N, 44°W). Note that towards the crest of the ridge, sediment cover is minimal or nonexistent but toward the abyssal plain side the basalt topography is being buried by detrital sediment. (Courtesy of Teledyne Exploration Company.)

The thickness of sediment along the major ocean ridge areas is generally small. It is not uncommon, however, to find thick ponded sediments in adjacent troughs and basins distributed as part of the rugged topography of this region of the oceans (Fig. 2-8). The ocean basin floor is composed of thick sediments, which generally obscure the topography in the basaltic rock beneath it.

ocean circulation

The oceans form a continuous body of water from the equator to the high latitudes. At the equator the direct rays of the sun provide the largest flux of heat energy per unit area on Earth. In the polar regions during the winter months, there is virtually no direct solar heat flux, but heat is supplied from low latitude regions by the movement of ocean waters and air masses. There is, of course, a return movement of cooled water and air as part of a circulation of worldwide dimensions, the driving force for which is the Sun's heat.

If we are to understand how the oceans circulate we must know how the physical properties of sea water are altered by heating and cooling, and how the planetary forces act on the oceans to influence their motions. In addition, since we are dealing with very large-scale features we must be able to keep track of the pathways of the ocean currents, and this requires suitable tracers of the water masses.

SAMPLING THE OCEANS

The two most useful measurements on sea water for dynamic studies are temperature and the dissolved salt content, and methods that give highly accurate results have been developed for obtaining them. Although both of these measurements can now be made by *in situ* probes in the ocean, it is still a major program in oceanography to obtain discrete samples of water for shipboard and shore-based laboratory analysis. The reasons for this are twofold. First, the *in situ* probes must be continuously calibrated and second, diagnostic chemical measurements, in addition to the total dissolved salt, are needed for the determination of rates of ocean circulation.

It is not a difficult matter to sample the surface of the ocean for chemical or physical analysis. One can dip a bucket or use a pump and hose and have all the water necessary for making the desired chemical or physical measurements. But it is a little harder to obtain a sample of water from the ocean deeps with accurate measures of depth and temperature and with assurance that the sample of water, once collected, is unmixed with surrounding waters during its return to the surface. An effective solution to this problem of sampling ocean water and ocean temperature at great depths was made by Fridtjof Nansen, the Norwegian oceanographer, in 1904, after whom a commonly used sampler is named. The bottles are spaced along a wire as it is let out from a ship by means of a metered winch (a device used to raise and lower cables aboard the ship). Hence, the depth of each bottle on the wire can be approximately determined. A more precise depth is obtained by measuring the amount of compression of a mercury thermometer in direct contact with sea water. The increasing pressure of the water with depth produces a different reading from that of a thermometer protected from the pressure effect. Thus, both temperature and depth can be measured with these two thermometers attached to the bottle. Each Nansen bottle samples about one to two liters of sea water. Most of the hydrographic data obtained to delineate the major features of the oceans were obtained by this method.

FIG. 3-1 A "rosette" of Niskin sampling bottles in two tiers. In addition to the bottles, an array of instruments including a conductivity-temperature-depth (CTD) probe, an oxygen probe and an instrument for measuring particle concentration (laser nephelometer) is usually mounted on the rosette. These instruments telemeter their information through the wire on which the rosette is lowered to a tape recorder or computer aboard ship. A signal from the control console aboard the ship can trigger the bottles to close, each of them at the desired depth.

Newer sampling techniques have been developed that take advantage of remote electronic manipulating and monitoring capabilities (Fig. 3-1).

CHEMICAL AND PHYSICAL PROPERTIES OF OCEAN WATER

Composition

If you were to sample the ocean all over the world with the above techniques and to use the most accurate methods of chemical analysis for determining the composition of the sea water at each location, you would find that, although the total amount of dissolved salts is variable, the relative proportions of the major elements (such as sodium, chlorine, magnesium, calcium) are constant. This fact was ascertained by successively more accurate analyses by chemists throughout the nineteenth century, culminating in the definitive study by W. Dittmar on water samples collected on the *Challenger* expedition in the 1870's.

The saltiness of the ocean, or *salinity*, is defined as the number of grams of dissolved inorganic salts in 1,000 grams of sea water. The total range of salinity of the open ocean is from 33 to 38 parts per thousand. The variations of salinity in the open ocean are the result of a number of competing processes: concentration effects, such as evaporation and ice-flow formation; and dilution effects, such as atmospheric precipitation, stream runoff, and melting ice.

Table 3-1 Concentrations of the Major Components of Sea Water*

Component	Concentration (grams per kilogram)
CHLORIDE	19.353
SODIUM	10.76
SULFATE	2.712
MAGNESIUM	1.294
CALCIUM	0.413
POTASSIUM	0.387
BICARBONATE	0.142
BROMIDE	0.067
STRONTIUM	0.008
BORON	0.004
FLUORIDE	0.001

* For a salinity of 35 parts per thousand; defined as the mass in grams of the dissolved inorganic matter in 1,000 grams of sea water after all bromide and iodide have been replaced with chloride, and all bicarbonate and carbonate converted to oxide.

Culkin, 1965, in Chemical Oceanography. Riley and Skirrow, eds.

Nearshore waters range both down to even lower salinities because of dilution by fresh water from streams and up to higher salinities because of intense evaporation in arid climates.

The average concentration of the main components of sea water with a salinity of 35 parts per thousand is shown in Table 3-1. Because of the constancy of the different major ions relative to each other in sea water, it is evident that the measure of the concentration of any one of these in a seawater sample would be an index of the salinity. It was common practice in the past to determine salinity by measuring the amount of chloride by chemical means. The relation between this measurement and the salinity is:

$$\text{salinity} = 1.80655 \times \text{chlorinity}$$

(chlorinity = grams of chloride equivalent in 1,000 grams of sea water)

Conductometric measurements of salinity are now commonly made either with remote probes sent down the wire with a pressure and temperature sensor or on seawater samples in the shipboard or shore laboratories. This type of measurement depends on the fact that a salt solution will conduct an electric current. At a given temperature the higher the salinity the greater the conductivity measured. Figure 3-2 shows a typical salinity profile in the ocean.

FIG. 3-2 Salinity and potential temperature profile in the North Atlantic (34°46.5′N, 67°59.8′W) showing the deep cold water, an intermediate depth water, and the thermocline and the halocline. The mixed layer is about 25 meters deep in this section and is not well represented. (After D. W. Spencer, 1972.)

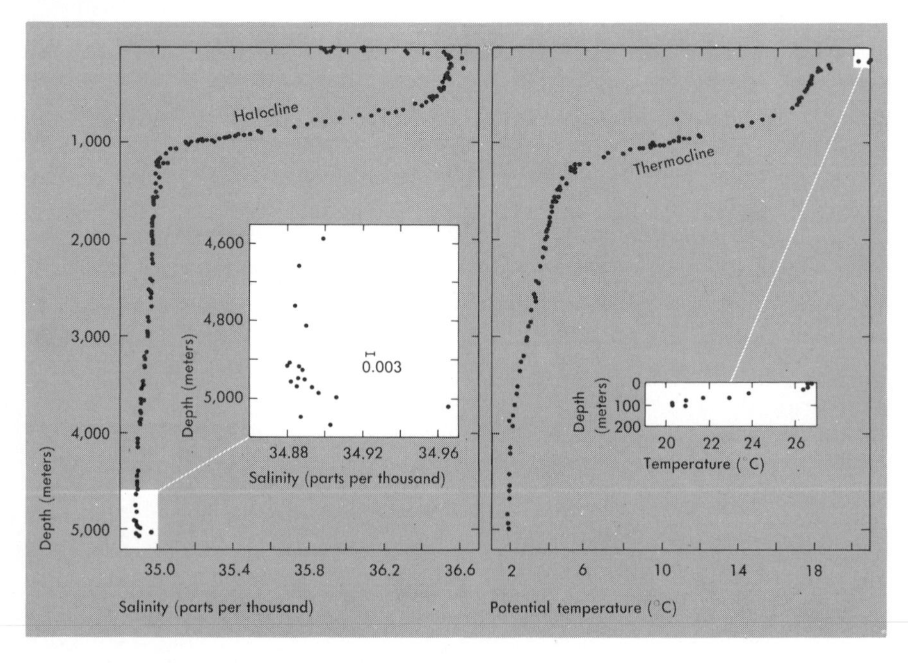

Temperature

We can think of the oceans as a gigantic pump that transfers heat from the equator to the poles. (It supplements the major transport of heat from low to high latitudes by the atmosphere.) This transfer is effected in the surface waters of the ocean by strong currents, such as the Gulf Stream moving warm tropical waters to polar regions. The deep waters of the oceans have their origins in the high latitudes. Hence, the deep waters are considerably colder than the surface water.

When water is subjected to the increasing pressure of depth it is compressed and becomes warmer, much like the heating of a tire during rapid inflation. In both cases heat is generated by the compression. In the case of a tire, it cools off again to the ambient air temperature because of circulating air about it. When the heat is not allowed to escape from the system, the process is called *adiabatic*, and compression must result in a rise in temperature. The temperature measured by an unprotected reversing thermometer or a temperature probe is, of course, the *in situ* temperature. Knowing the behavior of water under compression (that is, knowing the equation of state of sea water), we can correct the *in situ* temperature to the temperature the water parcel would have at the ocean surface. Since all water at one time was at the surface, this is a reasonable correction. The pressure-corrected temperature is called the *potential temperature* and is a significant parameter in determining the stratification in the ocean.

Figure 3-2 shows a typical temperature profile in the oceans. The main features of this profile are: (1) a surface, or *mixed*, layer reflecting the temperature of the ambient average temperature of that latitude, (2) a deep (and bottom) layer reflecting the origin of the water in high latitudes, and (3) a *thermocline* layer between about 100 meters and 1,500 meters in which the temperature generally decreases from the high surface value to the low deep value.

The thermocline layer indicates that there is transfer of heat vertically as well as horizontally from the surface waters to the deep waters. Although some of this transfer occurs by molecular heat diffusion, much of it is accomplished by small eddy currents that transport water vertically, thus mixing salinities as well as temperatures.

Density

When sea water is cooled its behavior is different from that of pure water (described in Chapter 1). The density of sea water increases continuously as it is cooled (Fig. 3-3). Thus as the ocean surface is cooled, the surface water tends to sink and displace deeper water until the whole system approaches the freezing point (if the system were virtually closed). As further cooling continues at the surface, sea ice is formed. This ice would be made of pure water except that salt grains are occluded during rapid freezing. It is still a fairly "fresh" ice with the result that the remaining sea water from which it formed increases in salinity. A higher salinity for a given temperature means denser water so that fingers of

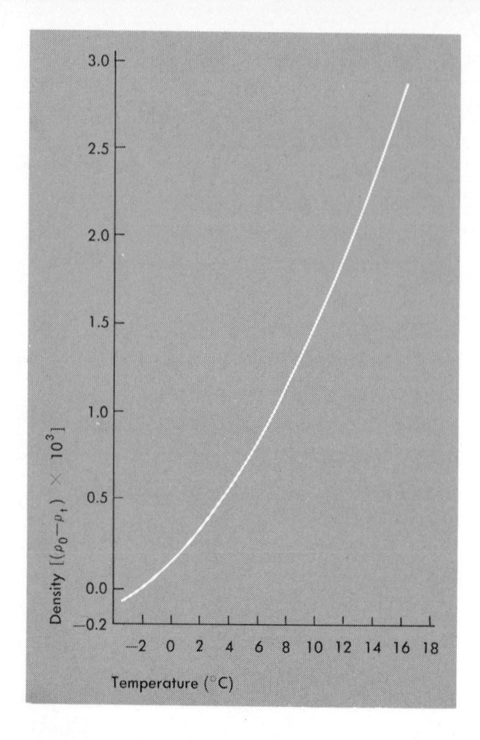

FIG. 3-3 (Left) The density of sea water with 35 parts per thousand salinity increases with decreasing temperature. ρ_0 is the density at 0°C and ρ_t is the density at t°C. (After Neumann and Pierson, 1966.)

FIG. 3-4 (Below) The relation between temperature and salinity below 200 meters for the Atlantic, Pacific, and Indian Oceans. The symbol σ_t represents the density minus one times 1,000 at one atmosphere pressure. A sea water density of 1.0280 thus has a $\sigma_t = 28.0$. The high-temperature–high-salinity region is the water at about 200 meters. The areas represent data for each total ocean system. There are distinctive water masses defined within each ocean as well. Note the effect of high-salinity water types injected into the Atlantic Ocean (Mediterranean water) and the Indian Ocean (Red Sea water). Antarctic Bottom Water, common to all the oceans, is the densest of all water masses. (After Dietrich, 1963.)

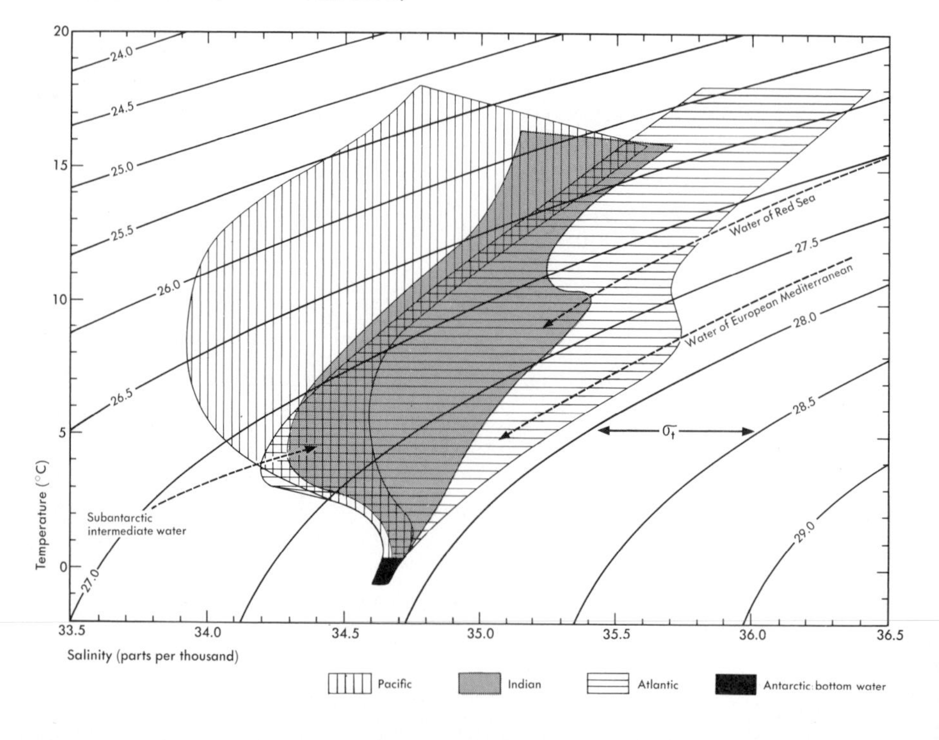

cold brine make their way to depth, displacing the less saline water to the surface from which more ice is formed.

This is one important method of modifying the density of sea water. Other processes, such as freshwater mixing, evaporation, or the melting of sea ice into the ocean water, will alter the salinity and temperature of a mass of sea water. The density of sea water is a function of temperature and salinity. The higher the temperature, the lower the density for a given salinity; and the higher the salinity, the higher the density for a given temperature. Figure 3-4 is a plot of salinity against temperature. Lines connecting points of equal density waters are called *isopycnal* lines. For example, it is evident that cold, highly saline water is denser than warm, low salinity water. However, any combination of salinity and temperature of one water parcel that yields a lower density than another parcel with a given salinity and temperature will float on top of this denser fluid. This operates everywhere in the ocean, resulting in a density stratification. Although ocean water of one density is completely miscible with ocean water of another density, the process of mixing takes a long time, since it depends on diffusion and convective mixing; hence waters of different densities tend to be stratified as long as water type supply rates are fast enough.

THE BEHAVIOR OF FLUIDS ON A ROTATING PLANET

To understand the reason for the circulation patterns in the ocean, we must first look at the laws affecting the behavior of fluids in motion on a rotating planet.

The Foucault Pendulum and the Concept of Vorticity

In 1851, J. L. B. Foucault demonstrated, without recourse to astronomic observations, that the Earth indeed does rotate on its axis. He suspended a pendulum in the Pantheon in Paris and carefully started it swinging in a plane whose intersection with the floor he marked with the time of the beginning of the experiment. He noted that the plane of the swinging pendulum rotated regularly and always took 32 hours to complete a cycle. Others have performed the experiment and have shown that the length of time to complete the cycle is 24 hours at the poles increasing to infinite time (no rotation) at the equator.

The explanation for the behavior of the Foucault pendulum is the following: According to Newton's laws of motion, the pendulum is swinging in a plane in the universe and will maintain that plane if no force is applied to turn it. Since we assume a frictionless attachment to the ceiling, no torque exists on Earth to rotate it. However, the Earth is rotating. Thus a pendulum located at the North Pole, while retaining its plane of rotation, will appear to rotate as the Earth rotates beneath it on a 24-hour basis. At the equator the plane of the equator and the plane of the pendulum will not change and therefore no movement of the swinging plane of the pendulum will be observed. The latitudes between 0 degrees (the equator) and 90 degrees (the poles) will have periods of rotation of the plane obeying the following law:

$$P = \frac{24}{\sin \theta}$$

where P is the period in hours and θ is the latitude.

Moving oceanwater parcels at different latitudes have behaviors resembling Foucault pendulums operating at these latitudes. The natural period of rotation of the parcel of water, or its *vorticity*, will be determined by the latitude: a larger vorticity at high latitudes with decreasing values to the equator.

The Coriolis Effect

Another effect that can be explained as the result of the Earth's rotation on a moving object is called the Coriolis effect after its discoverer, Gaspard de Coriolis.

If we looked down from the North Pole we would note that Earth is rotating in a counterclockwise sense. Suppose we are on a circular platform in space rotating in a counterclockwise sense and set up targets on the circumference. We then aim at the target and shoot a projectile. The projectile will appear to swerve to the right of the target as it continues on true course in the universal framework. This is the Coriolis effect: An object in motion in the Northern Hemisphere will be deflected to the right and in the Southern Hemisphere to the left. As with the law governing the rotation period of the pendulum, the extent of deflection is determined by the latitude. The deflecting "force" is greatest at the poles and least at the equator.

Western Boundary Intensification

One last effect that is due to the rotation of the Earth and that controls the behavior of ocean currents is the fact that the ocean basins have continental boundaries. The continents are rigid and the water is fluid. As a result the rotation of the Earth results in a tendency to pile up waters on the western boundaries of the ocean basins. This would not be the case if the Earth were not rotating in the direction that it does.

SURFACE CURRENTS: THE WIND-DRIVEN CIRCULATION SYSTEM

Figure 3-5 is a chart of the surface currents of the oceans. It is evident that the major features in the Atlantic and Pacific oceans are large circular patterns of currents ("gyres"), clockwise in the Northern Hemisphere and counterclockwise in the Southern.

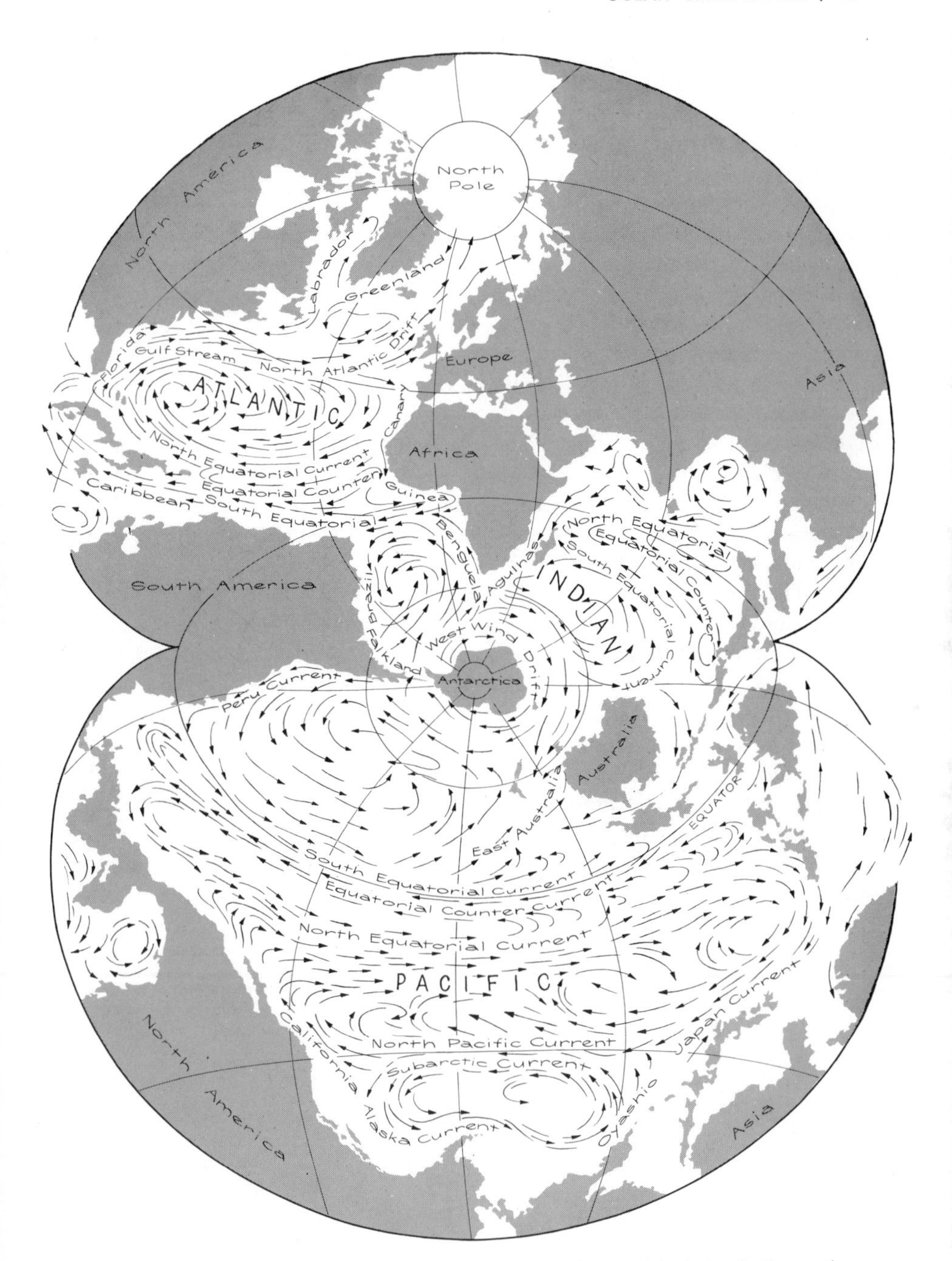

FIG. 3-5 The surface currents of the oceans. The pattern of gyres (clockwise in the northern hemisphere and counterclockwise in the southern hemisphere) can be explained as the result of the major global wind patterns—the prevailing westerlies blowing from west to east at about 40°N and 40°S and the trade winds blowing from east to west just north and south of the equator. (From Munk, 1955.)

The Gulf Stream in the North Atlantic is perhaps the most famous of the surface currents. It is part of the clockwise gyre system of the North Atlantic. Originating in the eastern end of the Gulf of Mexico, it moves northward through the Florida Straits hugging the continental margin as far as Cape Hatteras; it then turns toward the open ocean and, farther north and east, it becomes the North Atlantic Current (or "Drift") that is responsible for the warming of the British Isles. The Gulf Stream marks the boundary of warm surface water to the south and cold surface water from the north. At this boundary a strong narrow current develops that results in the transfer of warm water northward. It is to be noted that the Gulf Stream is not a current of warm water flowing *through* cold water.

To explain the surface circulation of the oceans we must start with winds. As heated air from the tropics is transported toward the poles in gigantic convection cells, the rotation of the Earth impresses itself on the motion. One consequence of this is that a pattern of strong prevailing westerly winds encircles both the Northern and Southern Hemispheres of the globe at about 40 to 50 degrees latitude. This is balanced by a prevailing easterly wind pattern called the trade winds, which occur at about 20 degrees latitude.

These winds act on the surface of the ocean, but not simply by pushing the water in front of it as one might at first imagine. The Coriolis effect is operative on a global scale and the force on the water will result in a deflection of the water at an angle to the direction of the applied force of the blowing wind. In the Northern Hemisphere, as we have seen, this is to the right, and water in contact with a wind will move to the right of the wind direction. If we imagine the surface water column to be composed of horizontal planes of water, then each overlying plane in motion will cause the underlying plane to move to the right of its motion. This is called the Ekman spiral (Fig. 3-6). On the average the water column influenced by the wind will move in a path at right angles to the direction of the wind.

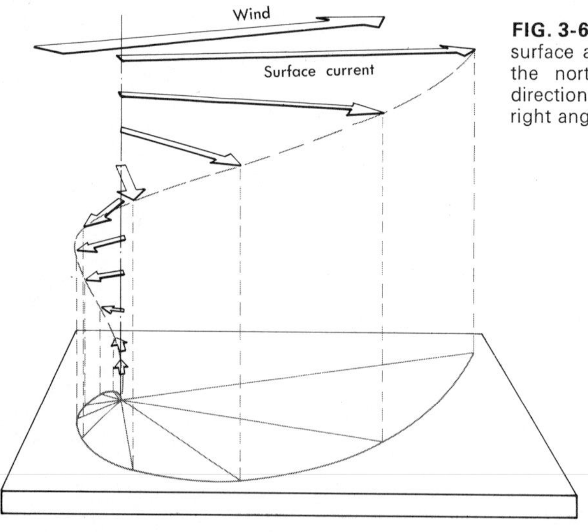

FIG. 3-6 The Ekman spiral at the ocean surface activated by the wind as seen in the northern hemisphere. The average direction for the column of water is at right angles to the direction of the wind.

This process will tend toward a piling up of surface water at the center of each hemispheric ocean basin. Of course, water cannot pile up because a pressure "head" will develop. The pileup at the surface causes water at a depth of 100 to 200 meters to be distended horizontally by squeezing. The water had taken on the rotation characteristics (the vorticity) of the particular latitude in which it is found before distention. Distending a layer of water causes it to lose vorticity due to conservation of angular momentum requirements. This water then moves toward the equator where the vorticity is lower so it can conform to its proper place on the rotating planet. As it moves towards the equator, it is deflected as a body to the right in the Northern Hemisphere, and thus to the west. As water has been lost from the gyre system by this southward transport, mainly from the region at 40 degrees to 50 degrees latitude, a return flow is required. The water supplied southward from the center of the gyre displaces surface waters, which flow with intensification along the western boundary to replace the missing water. This intense flow of warm water from the south is called the Gulf Stream in the Atlantic Ocean.

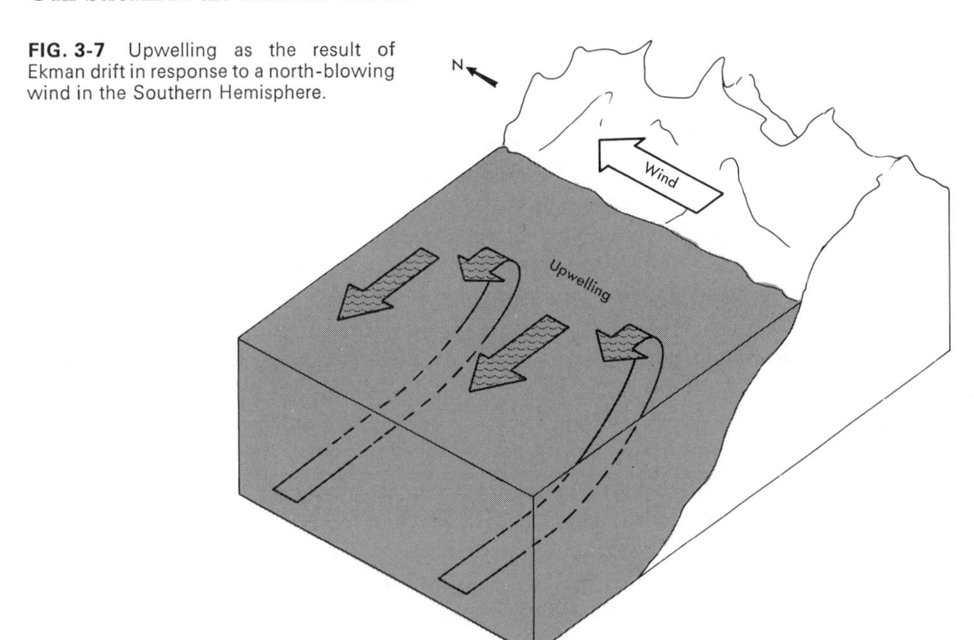

FIG. 3-7 Upwelling as the result of Ekman drift in response to a north-blowing wind in the Southern Hemisphere.

Along the western coasts of continents, surface water is transported away from the continent by north-south blowing, topographically intensified winds (Fig. 3-7). This is due to the Ekman spiral discussed earlier. This westward transport of surface waters results in the upwelling of deeper waters (down to depths of about 300 meters). As we shall see later, the deep waters are the major repositories of nutrient elements such as nitrogen and phosphorus. Areas of upwelling are thus marked at the surface by high biological productivity because of the continuous supply of these nutrient elements brought to the surface. Any change in the intensity of winds affects the rate of upwelling, resulting in dramatic changes in biological productivity.

Surface currents can easily be measured and their velocities determined by a number of techniques. One obvious technique is to track floating objects, be they ships or corked bottles with dated notes in them. On the basis of these observations we note that the surface oceans circulate completely in less than 10 years within each gyre and a little longer between gyres in the same ocean. As the turbulent regime of the Antarctic Ocean churns up the water to virtually all depths, there is no clear identification of surface water transport between the Atlantic and the Pacific. There is some flow, however, between the Indian and the Atlantic around the Cape of Good Hope and between the Indian and the Pacific through the passages around Indonesia and Australia.

THE DEEP CIRCULATION OF THE OCEANS: THERMOHALINE CIRCULATION

In deep circulation the controlling factor is density stratification in the Earth's gravity field. Water may become denser with increased salinity, decreased temperature, or both, hence circulation based on these factors is called *thermohaline*. The distinctive temperature and salinity attained by a body of water as a result of chilling, evaporation, freezing, dilution by rain or meltwater, or a combination of these processes will cause it to seek its level in the ocean based on its density. Since these distinctive waters are produced continuously from year to year, the oceans must keep circulating to accommodate them.

The oceanic structure of the Atlantic Ocean has been studied intensively through analyzing temperature and salinity profiles (Fig. 3-8). For an idea of the origin of various water masses, consider the following examples:

North Atlantic Deep Water (NADW) is believed to form when warm Gulf Stream water, which is highly saline as the result of evaporation in the low latitudes, is chilled by mixing with cold, less saline Arctic water near Greenland. The resulting dense water sinks to the bottom of the North Atlantic Ocean. From there it can be traced southward until it is obscured by mixing.

Antarctic Bottom Water (AABW) is formed when water in the Weddell Sea is chilled in the Antarctic winter so that relatively salt-free sea ice is formed, leaving a cold, very saline brine. This descends to the ocean bottom while mixing on the way down with less saline water. The mixing decreases its saltiness but not its temperature. What results is the densest water in the Atlantic Ocean and it can be traced along the bottom as far as 40 degrees north latitude.

The other water masses of Fig. 3-8 are formed in somewhat similar ways and occupy the levels dictated by their densities. All the water masses are the result of mixing of water types. The farther from the source of the water

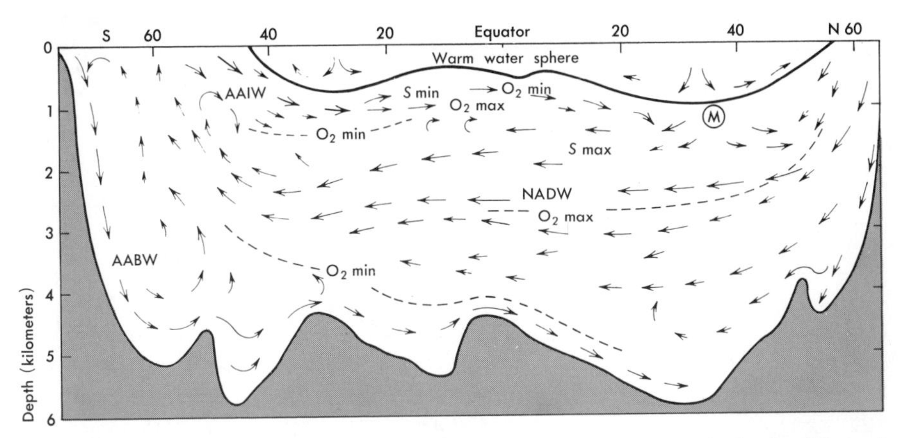

FIG. 3-8 The major water masses of the Atlantic Ocean identified on the basis of salinity and temperature. The arrows show the main directions of water flow. AAIW is Antarctic Intermediate Water and Ⓜ is Mediterranean water (flowing from east to west). The maximum and minimum oxygen concentrations are also shown, since they help to identify water masses. "S min" and "S max" are the salinity minimum and maximum layers in the deep ocean. (After Neumann and Pierson, 1966.)

the greater the amount of mixing with surrounding waters, until ultimately homogenization as the result of diffusion occurs; at this point the identity of the water mass is lost.

The general global circulation of the deep oceans, is represented in Fig. 3-9. The major sources of deep water are in the North Atlantic south of Greenland and several sites around the Antarctic continent. As the bottom waters flow away from their sources they are intensified at the western boundaries of the ocean basins. As water is supplied via these western boundary currents they spread out eastward and begin a return diffuse flow to the source areas. All along the way water is also advecting upward toward the ocean surface.

Cold dilute intermediate waters, formed by meltwaters mixing with surface waters in the circum-polar region, flow out and describe a counterclockwise gyre in the Southern Hemisphere and a clockwise gyre in the Northern Hemisphere. These waters are gradually blended into the surrounding waters at their farthest extensions.

If the deep circulation of the oceans ultimately depends on the flow of water formed at the high latitudes, there is obviously a horizontal bulk transport, or *horizontal advection*, of water. As this cold, dense water spread along the bottoms of the oceans it is also rising upward to replace the water transferred from the ocean basin to the high latitudes via shallower return flows. This results in *vertical advection* throughout the ocean basins. In addition to these advective modes of transport, water of one source gradually blends with surrounding water by a process that resembles *diffusion*—that is, the exchange of small parcels of water so that a mixing of properties occurs.

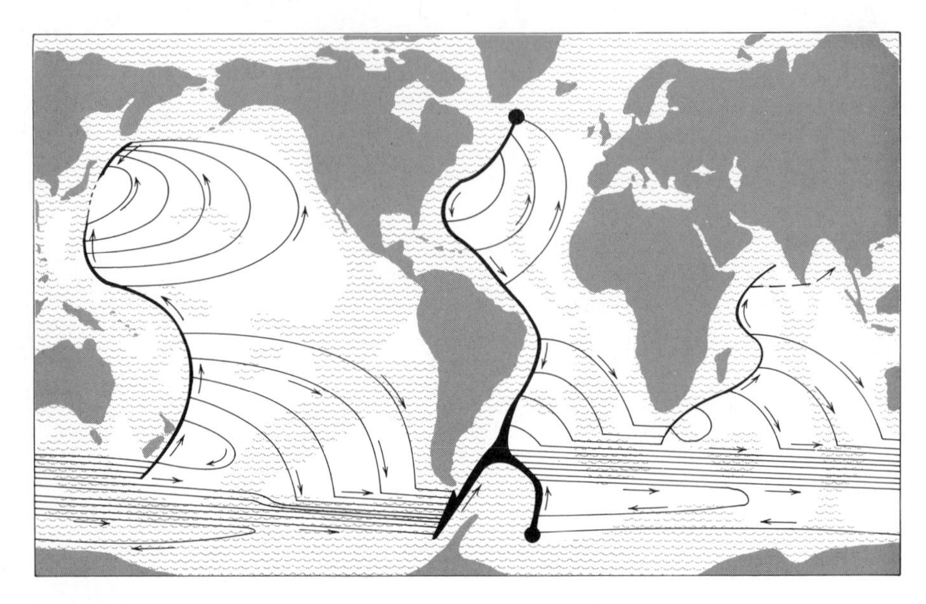

FIG. 3-9 Deep water circulation in the oceans. The horizontally shaded areas represent the major ridge systems of the oceans. The thick lines are the major bottom currents intensified along the western boundaries. The thin lines represent the "return" more general flow of water. (From Stommel, 1958.)

If, during these transport processes a property is altered on a systematic basis that cannot be explained by simple mixing, it is a *nonconservative* property and can be used to obtain a measure of the time involved in the transport process. There are two such classes of reactions: (1) the alteration of concentrations of elements involved in biological processes in the oceans such as the "nutrient" elements phosphorus, nitrogen, carbon, and silicon, and the metabolic "fuel," dissolved oxygen; and (2) radioactive clocks such as radium-226, carbon-14, and silicon-32.

Methods using all of these materials have been developed to assess the movement rates of water. Many of the measurements and calculations are continually being refined, but a good idea now exists about the time scales of deep-ocean circulation rates.

If we think of the oceans as big interconnected basins we can ask our question about circulation rates in terms of the average length of time that a water molecule spends in each deep basin. The results of such calculations indicate that water molecules spend from 200 to 500 years in the deep Atlantic before being transferred to another reservoir (in this case the deep Pacific via the Antarctic Ocean), and about 1,500 to 2,000 years in the deep Pacific before returning to the Antarctic Ocean region (via southward flows at shallower depths).

four

tides, waves, and the coastal ocean

In addition to the large-scale circulation of the oceans, there are other types of motions of the sea. These are commonly part of the experience of people living near the coast. The systematic rise and fall of sea level, which we call the tides; the beating of the surf—the death knell of waves; and the interplay of fresh water with salt water at the mouths of rivers: all of these are often our first encounter with the endless sea. The causes and effects of waves, tides, and coastal processes will be discussed in this chapter.

TIDES

The periodic rise and fall of sea level is called *tide*. It has been known for some time that there is a close relationship between the phases of the moon and the amplitude of tides.

Tides are not everywhere the same in amplitude. People living around the Mediterranean would hardly be aware of the phenomenon since there is virtually no tide in that sea. On the other hand, the coastal inhabitants of the North Sea, and especially Brittany in France, could not help but be impressed by the twice-daily flood and ebb of the oceans.

The forces producing tides are astronomical. The attractive pull of the Moon and to a lesser extent the Sun cause water to move with periodicity on the Earth's surface. If we consider the effect of the Moon on a completely water-covered Earth, we can understand the general mechanism of tidal forces and the effect under an idealized equilibrium situation. Referring to Fig. 4-1

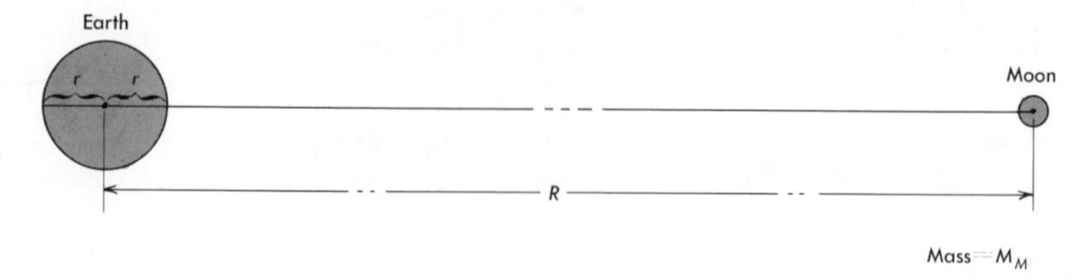

FIG. 4-1 The Earth-Moon system.

we designate the radius of the Earth as r; the distance between the center of the Earth and the Moon as R; and the mass of the Moon as M_M. The force due to the Earth-Moon system on a unit mass of Earth anywhere along the radius r would be the same if all parts of the Earth were rigidly joined to the center. This force is (see Chapter 1):

$$F_E = \frac{GM_M}{R^2}$$

The force on a unit mass on the surface of the Earth not anchored rigidly to the center of the planet (such as sea water) is:

$$F_W = \frac{GM_M}{(R - r)^2}$$

The difference in force between the rigid Earth and the nonrigid water is then:

$$\Delta F = F_W - F_E = GM_M \left[\frac{R^2 - (R^2 - 2rR + r^2)}{R^2(R - r)^2} \right]$$

$$= \frac{2r(R - r)}{R^2(R - r)^2} = \frac{2r}{R^2(R - r)}$$

Since $R \cong 600{,}000$ kilometers and $r \cong 6{,}000$ kilometers, R is clearly much greater than r and we can approximate $(R - r)$ by R, thus:

$$\Delta F \cong \frac{2r}{R^3}$$

The differential force between the rigid earth and the fluid water on the surface would tend to pull the water toward the Moon under the sublunar point. By exactly the same reasoning on the diametrically opposite side of the Earth, the Earth is being pulled differentially away from the surficial water. Thus water would bulge out at the sublunar point and also on the opposite side.

Of course no strong stretching of water is possible since the chemical bonds are too strong. Indeed the relatively strong incompressibility of liquids is the basis of operating hydraulic elevators. So actually the tide is not produced by the direct vertical pull of the Moon on a column of water.

Over the entire surface of our idealized water-covered planet the differential force varies from point to point. At the sublunar point and its diametrically opposite point the force is completely vertical, but over the rest of the Earth the force on water molecules has an increasingly strong horizontal component as we proceed away from the two points. It is this differential horizontal force distribution that causes water to flow on the Earth toward the sublunar point and its diametrically opposite point (Fig. 4-2).

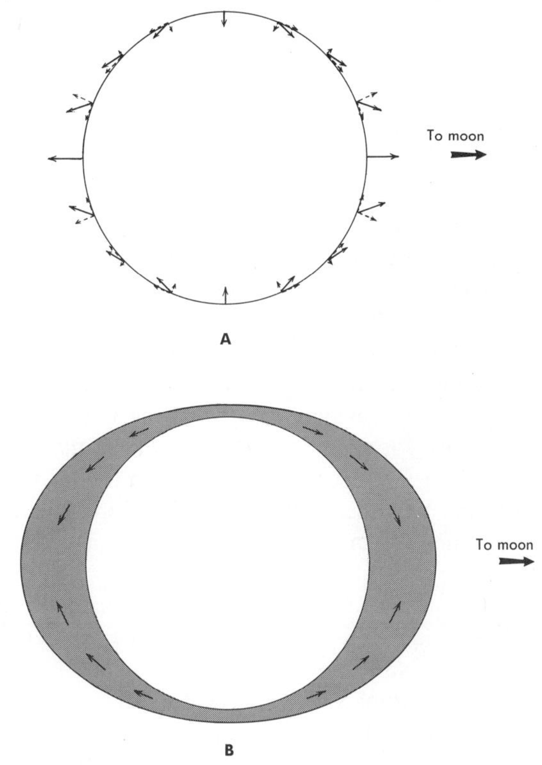

FIG. 4-2 The tide-producing force is a combination of the pull of the Moon (and Sun) and the pull of the Earth on the oceans. (A) A representation of the net force operating on water molecules on different points on the Earth's surface. (B) The resulting flow of the oceans on an idealized water-covered Earth.

A particular point on the Earth's surface will experience, in one day, two high tides and two low tides. This is called a semidiurnal tidal cycle. Since the solar day is 24 hours, whereas 24 hours and 50 minutes pass before the Moon is again over the same spot on the Earth (because of the Moon's revolution around the Earth), the beginning of a particular cycle at a point on the Earth will be offset 50 minutes each day from the previous day.

In addition to the dominant effect of the Moon, the Sun also exerts a strong tidal effect on the Earth (about 46 percent of the Moon's force). Since

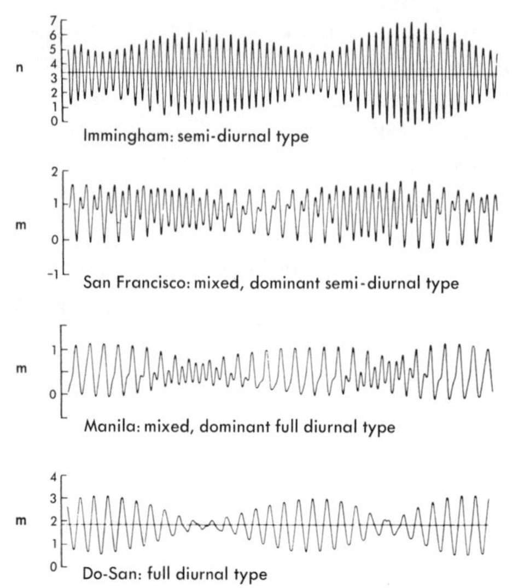

FIG. 4-3 Tidal patterns during March, 1936, at Immingham, England; San Francisco, California; Manila, Philippines; and Do San, Vietnam. The patterns differ because of the differing relative effects of the forces exerted by the Sun and Moon at the various locations on the Earth. (After Defant, 1961.)

the plane of the Moon's orbit around the Earth is not exactly in the same plane as the Earth's orbit around the Sun, there are complex effects on the tidal cycles generated by the action of the two forces. Hence, in different parts of the Earth one can find mixtures of semidiurnal and full diurnal tidal cycles, depending on the lunar and solar effects. These are seen in Fig. 4-3, which compares tidal gauge records for several different locations on the world ocean.

It is evident that when the Earth, Moon, and Sun are lined up, the tidal range for semidiurnal areas will be maximum because of the combined effects of the two astronomic bodies on the Earth. These occur at full and new moons and are called *spring tides*. When the Sun and Moon are at right angles to each other (at quadrature), the tides have the smallest amplitude and are called *neap tides*.

Since the tides are periodic phenomena, in a fluid they are sensitive to the boundaries of the container—the configuration of the oceans, bays, sounds, inlets, and so on. For this reason, at a particular location the tidal cycle may be very complex. A great deal of study has gone into constructing tide tables for seaports all over the world to predict the tidal sequences and ranges.

WAVES

Waves can be generated in a body of water such as a swimming pool or a lake by one of two methods. You can drop a pebble in the water and see a concentric wave pattern develop and spread away from the point of impact, or you can blow air over the surface of the water by means of an electric fan or other device. Similar processes are involved in making waves in the oceans. Although there is no significant horizontal transport of water, energy is transmitted through the wave action to areas remote from the point of disturbance.

The analogy to dropping a pebble in the water is the transmission of energy to the ocean by an earthquake, volcanic explosion, or landslide on the ocean's margins and floors. Such a wave is called a *tsunami* and travels at a velocity of 800 kilometers per hour. These are relatively rare events compared to the continuous wave generation by winds, but they are nevertheless quite spectacular and destructive when they encounter coasts.

By far the most common waves in the oceans are those generated by winds. When a wind blows over the surface of the oceans it piles up the water in ridges, whose height and periodicity reflect the intensity of the wind. If the oceans were initially glassy smooth and flat, a sustained wind of constant strength and direction would produce a clearly discernible rippled effect observable for great distances away from the point of generation. Of course, the interaction of winds with the ocean varies in both direction and intensity, resulting in a complex pattern called a "sea." In principle, however, such a complicated pattern can be resolved into a combination of regular wave patterns.

The depth in the oceans influenced by waves (the wave base) varies with the wave length. In the deep ocean the level of the wave base is shallow (less than 50 meters) relative to the depth of the ocean (about 5,000 meters). In shoaling waters, as the coast is approached, the wave base will eventually encounter the bottom and the properties of the wave will be influenced by this encounter.

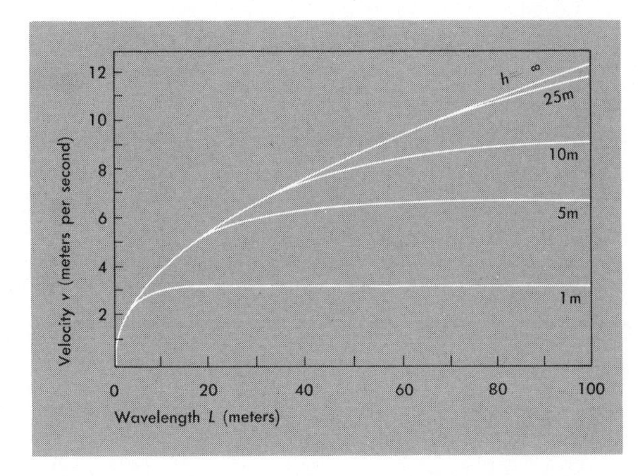

FIG. 4-4 Relation between wavelength and velocity for different depths (h) of water. (After Neumann and Pierson, 1966.)

On the basis of theoretical and experimental studies the relationships among wavelength, depth of water, and velocity of the wave front are well understood. Figure 4-4 shows the relationship between wavelength and velocity for different depths of water. It is obvious that in deep water (greater than 50 meters) the velocity of a wave is related only to wavelength. The relationship is:

$$v = \sqrt{\frac{gL}{2\pi}}$$

where v is the velocity, g is the acceleration of gravity (980 centimeters per second per second), and L is the wavelength.

In shoaling waters at depths less than 25 meters the speed of a wave is insensitive to wavelength but is strongly controlled by depth. The relationship is:

$$v = \sqrt{gh}$$

where v and g are as defined above and h is the depth of water.

The complex pattern of waves is called a *wave train*, and it moves away from the area of generation by wind with a "group" velocity. Within the wave train, however, the different wavelengths (or "phases") move with different velocities according to the equation for deep-water waves. It is evident that the longer the wavelength the greater the velocity. Thus the first arrivals of the waves generated by a storm far from the point of observation will be the long waves, or *swell*. The group velocity is about one half the phase velocity of the longest wavelength in the wave train. The shorter wavelength waves are absorbed in the longer waves along the way, and the final result is that remote areas receive only a continuous swell from the areas of storm disturbance.

In shoaling waters, the velocity of a wave is controlled by depth of water alone. The wave front (or the imaginary line delineating the major linear trend of the modified waves), then, replicates the submerged features of the coast, as discussed below.

The period of a deep-water wave is directly proportional to the velocity, which in turn is related to the wavelength. A wave with a long wavelength arrives at a high velocity toward the shoaling waters of a coastline and is slowed down.

There can be, however, no change of the wave's period in the new shallow water regime. Thus the wavelength of the incoming wave decreases as the water shoals and the velocity of the wave decreases. This results in an increase in the height of the wave. At a depth of water equal to about 1.3 times the wave height, the wave breaks. After the wave breaks near the strand line—the line where the sea meets the land—(in places the phenomenon of breaking occurs several times before the strand line is reached), the wave is expended on the shore as surf and swash.

THE COASTAL OCEAN

The boundary between the land and the sea is where some of the energy supplied to the sea by the Sun is dissipated. The Sun drives the winds to make the waves that impinge on the land. The Sun also causes water to evaporate from the sea, and this subsequently precipitates on land and channels its way back to the sea as rivers. Thus a significant amount of the energy supplied to the oceans by the Sun is dissipated along a narrow zone fringing the continents. The coast is also subject in many places to the influence of the astronomically controlled tides.

The Work of Waves

The refraction of waves approaching a submerged topography, as we have seen causes a shaping of the wave front to conform to this submerged topography. The result for a rugged hill and valley pattern is to cause the concentration of wave energy at the headlands (Fig. 4-5). This can be seen by drawing equally spaced lines perpendicular to the crests of waves. As long as the waves remain deep-water waves, the energy carried by each linear distance along the wave is the same. If we follow the originally equally spaced *orthogonals* (or rays), as the perpendicular lines are called, we note that as the water shoals they cluster together at the headlands and disperse in the bays. This results in a higher density of wave energy per unit coastline at the headlands than in the bay.

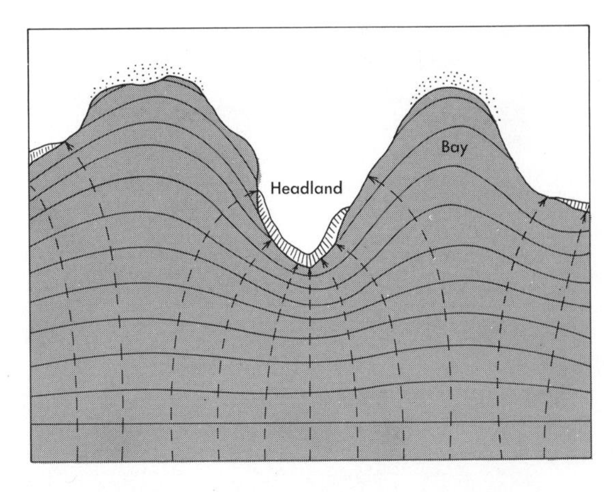

FIG. 4-5 The refraction of waves along a submerged coastline in which the bottom topography is a continuation of the adjacent land topography. The lines perpendicular to the wave fronts are called orthogonals and represent the distribution of energy along the coast. A tight spacing of the orthogonals means the dissipation of much energy along the coast with attendant erosion. Longshore currents transport this erosion debris to the protected bays.

The consequences of this in shaping the coastline are twofold: (1) the headlands are gradually eroded by the wave action, and (2) longshore currents develop away from the headlands toward the bay, causing the transport of fine debris to the shoreline along the bay. Eventually the bays will be filled with sediment and the headlands will be eroded. The underwater topography will also be filled in and a straight shoreline will evolve. The rate of this process clearly depends on the resistance to erosion of the rock or sediment making up the headlands. Essentially all coastlines have been submerged most recently about 11,000 years ago (due to the melting of glacial ice—Chapter 7).

The dominant direction of waves impinging on a shoreline will result in the development of longshore currents capable of transporting sediments. As this transport of sediments can seriously affect the quality of beaches, it has been common practice to retard it by building out *groins* that act as interceptors of the sediment being transported.

Where the River Meets the Sea

Rivers transport three major types of material to the ocean: (1) the water itself, (2) the dissolved burden of inorganic and organic substances, and (3) detritus such as organic material, sand, silt, and clay. Each of these influence, in some way, the properties of the coastal ocean. The fresh water flowing into salt water causes a circulation that we call *estuarine*; the dissolved load provides a supply of nutrient elements such as nitrogen and silicon; and the detritus helps to shape the coastline.

By the time a large river reaches the sea it has developed a high velocity and commonly, because of turbulent flow and the availability of erodable sediment along its course, it carries a sizable suspended load of clay, silt, and sand (see Table 5-1 in Chapter 5 for a summary of size classification). When the fast-flowing river encounters the large, relatively quiescent ocean, the sudden

FIG. 4-6 The Nile Delta as photographed from Gemini-4 (June 3–7, 1965). The Mediterranean Sea is to the left of the picture and the Suez Canal and Red Sea to the right. The rich soil of the Nile Delta is cultivated, thus giving it a dark appearance relative to the surrounding desert. (NASA photograph.)

change in velocity causes the sediment to drop out rapidly. In addition, the colloidal clay fraction is coagulated by the encounter with the salty water of the oceans and this further causes a settling out of sediment.

Rarely, as in the Congo River in Africa and Magdalena River in South America, where there is virtually no continental shelf, the river-borne sediment is dumped almost directly into the ocean depths via offshore canyons. Most commonly, however, rivers debouch onto a continental shelf. Depending on the strength of the forces of dispersal—mainly tidal action and storm waves and their associated longshore currents, sediment may either remain on the shelf at the mouth of the river or be dispersed on the shelf or farther seaward. If it accumulates at the mouth it is called a *delta*—named after the Greek letter that its aerial configuration resembles (Fig. 4-6).

Deltas occur most commonly where rivers empty into ocean basins that have very small tidal amplitudes, such as the Mediterranean (the Nile, Po, and Rhone deltas), the Gulf of Mexico (Mississippi delta), and the Arctic Ocean (Lena delta). Tides with large amplitudes act as strong dispersing forces. If the sediment load is overwhelmingly large, such as in the rivers draining the Himalayas, no dispersive force is great enough to carry away all the material that is transported by the rivers and dumped at the ocean boundary, and large deltas can develop.

As sediments accumulate on the shelf, they will eventually reach a point where they will descend through submarine canyons to form deeper continental margin deposits like deep-sea fans. Some fan building is active now but most fans were probably produced prior to the end of the latest worldwide glaciation, which started about 75,000 years ago and ended 11,000 years ago (see Chapter 7).

The supply of fresh water to a coastal area by streams strongly influences ocean circulation there. As the fresh water flows into the ocean it will tend to ride over the denser salty sea water, and in certain quiescent oceanic regions, such as some fjords, this indeed does happen. In most cases, the action of tides and waves mixes the fresh water and salt water. Since the stream water essentially is added to the top of the ocean, its tendency is to flow outward over the ocean water (Fig. 4-7). As the river plume mixes with the ocean a diluted sea water actually flows away from the river mouth at the surface. To balance the loss of sea water entrained with the outflowing riverine discharge, there must be a supply of sea water from the open ocean, which, because of its higher density, flows toward the river mouth at depth. This is estuarine circulation.

The effect of estuarine circulation on sediments is to keep pushing them back toward the continent. Only the finest particles that have not coagulated but have remained dispersed in the outgoing surface water make it out of an estuary. The tendency is for the sediments dropped by the river or eroded by waves to be pushed toward shore. Only when the pileup is sufficient to cause slumping or when intense longshore currents exist is there a good opportunity to lose sediments to the deep ocean from the shelf.

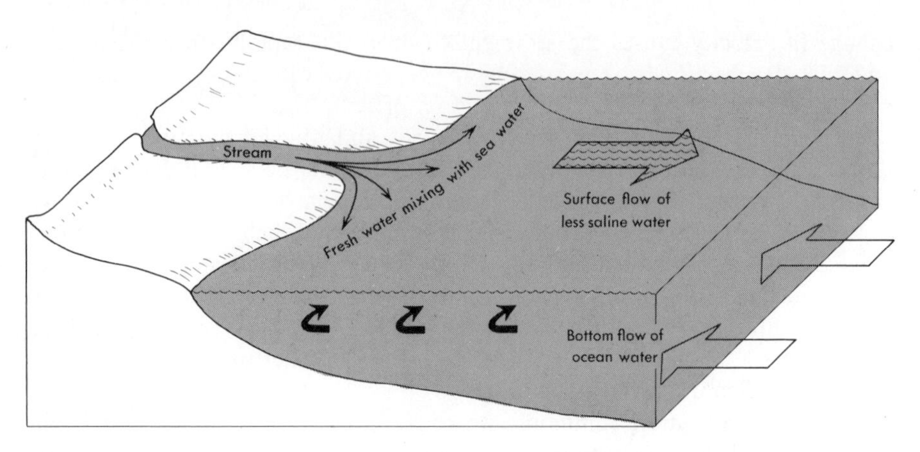

FIG. 4-7 Estuarine flow results from the mixing of fresh water from streams with sea water as a result of tidal and wave mixing. The mixed water flows out on the surface and sea water is returned at the bottom to make up for the sea water that is transported away from the coast at the surface.

Most estuarine circulation is identified with submerged river valleys or other undulating features, such as the Thames estuary in England or the Chesapeake Bay estuary bounded by Maryland and Virginia. Estuarine circulation with its attendant transport features exists, however, wherever there exists a significant flow of fresh water from the continents to the ocean.

sediments: sources and transport

Speaking in the broadest possible terms, the beginning point of the oceanic deposition of sediments begins on land and runs as follows: Rocks exposed above sea level become subject to chemical action, especially chemical influences associated with life; this process is called *weathering*. Some elements, leached from the rocks by weathering, are carried to the sea in solution; much of the remaining altered rock material becomes a complex of degradation products, such as clay and sand, characteristic of the local climate and general environment. These products of weathering are carried, by streams primarily, to the sea, together with particles from easily eroded rocks, which may not be strongly altered chemically. Material may also be supplied to the ocean basins by mechanisms other than streams that drain the continents. The products of oceanic volcanoes or of continental glaciers are also found on the ocean floor. In this chapter, we discuss the composition of sediments and the transport of particles both to and within the oceans.

WEATHERING, EROSION, AND THE DENUDATION OF THE CONTINENTS

When plants decay through the action of bacteria, carbon dioxide is produced as a metabolic product. Carbon dioxide will react with water to form carbonic acid. Waters charged with carbonic acid percolate through the rocks on which the plants grow, and attack the rock. This is the process of chemical weathering induced biologically. (In addition to carbonic acid, certain organic acids produced by bacteria also act as agents for the destruction of rocks.)

An idealized chemical reaction describing this weathering process can be written:

$$\text{Na feldspar} + H_2CO_3 = Na^+ \text{ (aqueous)} + HCO_3^- \text{ (aqueous)} + SiO_2 \text{ (aqueous)} + \text{kaolinite}$$

where:

Na feldspar is a representative of minerals in common rocks such as granite and basalt;

H_2CO_3 is carbonic acid formed by the charging of water with metabolically produced carbon dioxide;

Na^+ *(aqueous)*, HCO_3^- *(aqueous) and* SiO_2 *(aqueous)* are the dissolved products of the reaction; and

kaolinite is a representative of the type of residual degradation minerals (see next section).

The rates of chemical weathering have been measured by assessing the total burden of dissolved substances in a stream draining a known area and calculating the amount of rock that must be altered in this area per unit time to provide the measured flux. Such a study done in the Sierra Nevadas indicates that these mountains are decreasing in elevation by this chemical process at about 0.3 centimeters per 1,000 years.

Land actually is lowered faster than this because much of the degradation is due to physical erosion. That is, loose material from a soil profile or from old sedimentary strata may be eroded under the action of water, much as a pile of dirt can be washed away with a hose.

We can measure the sediment flux from the weathering and erosion of land by measuring the flux along stream-gauging stations or estimating the pile of debris supplied by streams that have come to rest in the ocean floor as deltas, deep-sea fans, and abyssal plains.

If we combine all the information available from data on streams we find that about 3.6×10^{16} liters per year of water enter into the oceans. The discharge per unit area of the oceans is thus about 10 liters per square centimeter per 1,000 years. The average dissolved load of streams is about 100 parts per million (see Table 9-4 for composition). Thus almost 1 gram per square centimeter per 1,000 years of dissolved salts are supplied to the ocean basins. The average particulate load is about 500 parts per million, thus about 5 grams per square centimeter per 1,000 years.

The corresponding rate of lowering of the continents by erosion, if no further mountain building elevates parts of them again (as must of course be the case), is about 6 centimeters per 1,000 years. Since the average elevation of the continents is 800 meters, it would take about 13 million years to lower the continents to sea level. We have geological evidence of lands and mountains for billions of years, so we conclude that the continents are renewed by mountain building and continental uplift fast enough to keep up with the erosion rate. (The significance of this will be shown in Chapter 8.)

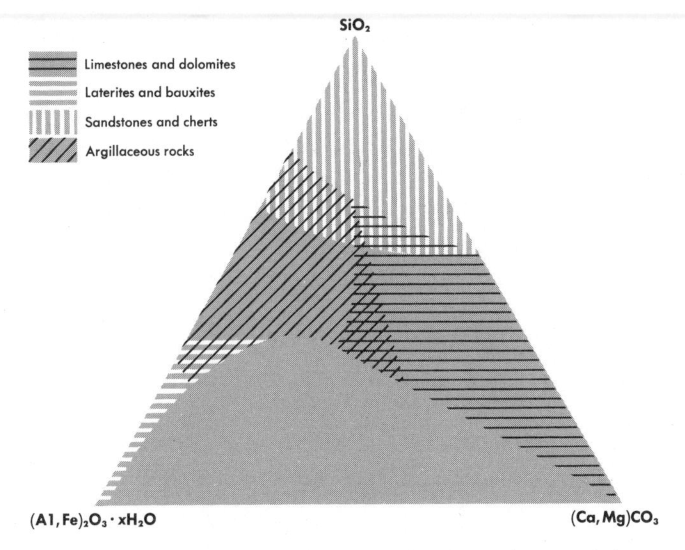

FIG. 5-1 The range of sedimentary rock types represented as mixtures of three components: calcium (plus magnesium) carbonates, clay minerals (represented by the hypothetical hydrated aluminum and iron oxides as the end member), and silica (silicon dioxide). Sediments and sedimentary rocks have the same ranges of composition. Iron-rich laterites and aluminum-rich bauxites are the products of intense weathering. Sandstones are primarily composed of indurated sandy sediments, in many cases dominantly quartz. Cherts are the sedimentary rock equivalent of biologically deposited siliceous deposits. During the transformation into rock, the amorphous silica originally deposited by diatoms and radiolarians is transformed into a very hard microcrystalline quartz-rich rock. Argillaceous (from the French *argile*, for clay) rocks are derived from the lithification of clay-rich muds. Sediments or sedimentary rocks rarely, if ever, have compositions represented by the unpatterned area of the triangle. (After Mason, 1967.)

THE COMPOSITION OF SEDIMENTS

Marine sediments are composed of detrital material from land and substances extracted from solution by biological or chemical processes. There are two gross classifications we can apply to marine sediments without being concerned about how a particular sediment got where it did. One is based on the grain size of the sediment (Table 5-1), the other on its composition (Fig. 5-1). If a sediment on the ocean floor contains particles nearly all of one size range, the sediment is *well-sorted*. Similarly, a sediment composed of one mineralogical or chemical type is *very pure*. Most sediments, however, are neither

Table 5-1 Sizes of Sedimentary Components*

Name	Particle Diameter (millimeters)
BOULDERS	Greater than 256
COBBLES	64 to 256
PEBBLES	4 to 64
GRANULES	2 to 4
SAND	0.062 to 2
SILT	0.004 to 0.062
CLAY	Less than 0.004

* Commonly called the "Wentworth scale."

perfectly well-sorted nor very pure; but it is precisely the information contained in the mixtures that is of great value in determining the history of the sediment. Of particular interest, because of their extensive distributions, are the clay- and sand-size components. The mineralogical composition in these size ranges varies considerably, depending on the sources of material.

FIG. 5-2 Diagrams looking at right angles to the sheets making up the main structure of the clay minerals. The silica and associated layers are stacked along the *c*-axis. The kaolinite sheets are held together by weak hydrogen bonds. Montmorillonite as shown is representative of a group of similar minerals in which substitutions of iron and magnesium occur at various sites. The iron-rich montmorillonite is the dominant mineral of deep-sea clays of the South Pacific. They share the property of holding water molecules between the sheets, causing expansion and contraction along the *c*-axis during hydration and dessication, respectively. The montmorillonite minerals also show a high capacity for cation exchange. *Illite* is the term used for the sedimentary fine-grained equivalent of ordinary mica (muscovite). The chlorite crystal represented here can be modified into an iron-rich form. Deep-sea sediments commonly contain this more iron-rich form. (After Mason, 1967.)

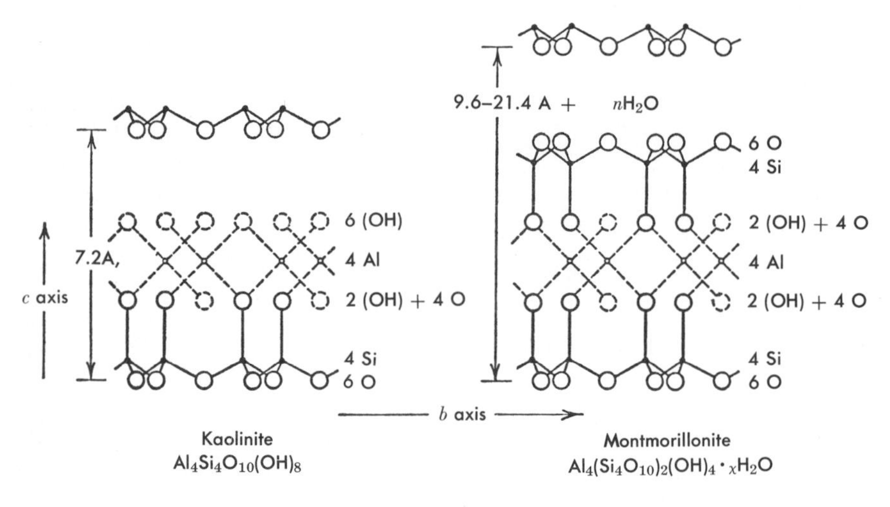

Kaolinite
$Al_4Si_4O_{10}(OH)_8$

Montmorillonite
$Al_4(Si_4O_{10})_2(OH)_4 \cdot xH_2O$

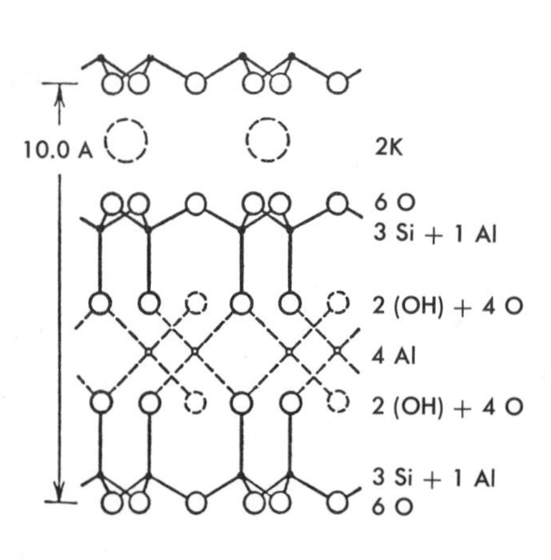

Illite (Muscovite)
$K_2Al_4(Si_6Al_2)O_{20}(OH)_4$

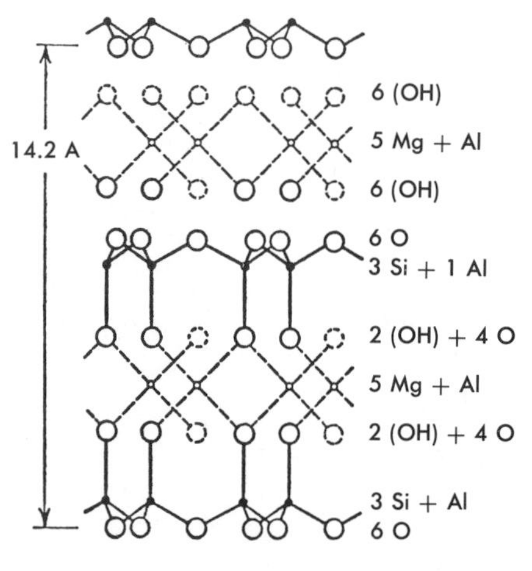

Chlorite
$Mg_{10}Al_2(Si_6Al_2)O_{20}(OH)_{16}$

Clay Minerals

Although the clay-size fraction (or portion of sediments) may contain fine-grained calcium carbonate, it is composed mainly of silicate and oxide minerals. The most important of these are the *clay minerals*, which are the silicate minerals formed from normal rocks either by weathering or the action of hot water from volcanic and other thermal sources. The principle clay minerals are kaolinite, montmorillonite, illite (or mica), and chlorite. The clay minerals, structurally, are related to the common mineral mica in that sheets, formed by the joining together of silica tetrahedra in a two-dimensional array, constitute the basic structural units (Fig. 5-2). The deviations from the mica structure and the variations among clay minerals are due to the way the silica sheets are stacked with other chemical layers and the degree of chemical substitution within both the original silica sheet and the added layers. Most of the clay minerals found in sediments are derived from weathering profiles on the continents and alteration products of volcanic rocks. They may be modified to some degree by interactions with sea water.

Calcium Carbonate and Related Minerals

Most of the carbonate minerals deposited in the oceans are due to the action of organisms. Calcium carbonate in one of three forms is the primary deposit: *Aragonite* is laid down by most present-day corals and some mollusks

FIG. 5-3 Aragonite-depositing marine organisms. (Top) Coral. All common reef-building corals deposit pure aragonite shells. This is the common "lace coral" (diameter about 20 centimeters). (Bottom) Mollusk. Both snails and clams have species that deposit aragonite in various amounts. This marine snail, *Conusgloria maris* (about 5 centimeters long), deposits a pure aragonitic shell, as does the chambered nautilus, a representative of another type of mollusk, related to the octopus and the squid. (Both courtesy American Museum of Natural History.)

(Fig. 5-3); *low-magnesium calcite* is deposited by some mollusks, some foraminifera (including all the deep-sea species), brachiopods, and the unicellular plant family *Coccolithophoridae* (a major constituent of deep-sea sediments and the main component of the chalks of the White Cliffs of Dover) (Fig. 5-4); and *high-magnesium calcite* is derived from echinoderms and some large foraminifera in shallow waters (Fig. 5-5). Only the mollusks commonly have species whose shells (or *tests*) are layers of aragonite and low-magnesium calcite.

In the geologic record, the older the rocks the less well preserved aragonite is, because it is not the stable state of calcium carbonate at low pressures. It is commonly replaced by calcite, even in some relatively young coral reefs now exposed above sea level, as in the Florida Keys.

FIG. 5-4 Low-magnesium calcite-depositing marine organisms. (A) Mollusk. This is the common oyster which deposits a pure low-magnesium calcite shell. (Courtesy Carolina Biological Supply Company.) (B) Pelagic foraminiferan. This test (*Globigerina bulloides*) is from the Scotian Shelf. Magnification 195 times. (Courtesy G. A. Bartlett, Bedford Institute of Oceanography.) (C) Coccolithophore. This is a shell deposited by a common type of marine plant (phytoplankton). The species is *Coccolithus huxleyi*. This coccosphere disaggregates during deposition and is rarely preserved intact. The component "ovals" and smaller platelets are common components of deep-sea sediments and many chalk deposits like the White Cliffs of Dover. Magnification 1,035 times. (Courtesy A. McIntyre, Lamont-Doherty Geological Observatory.)

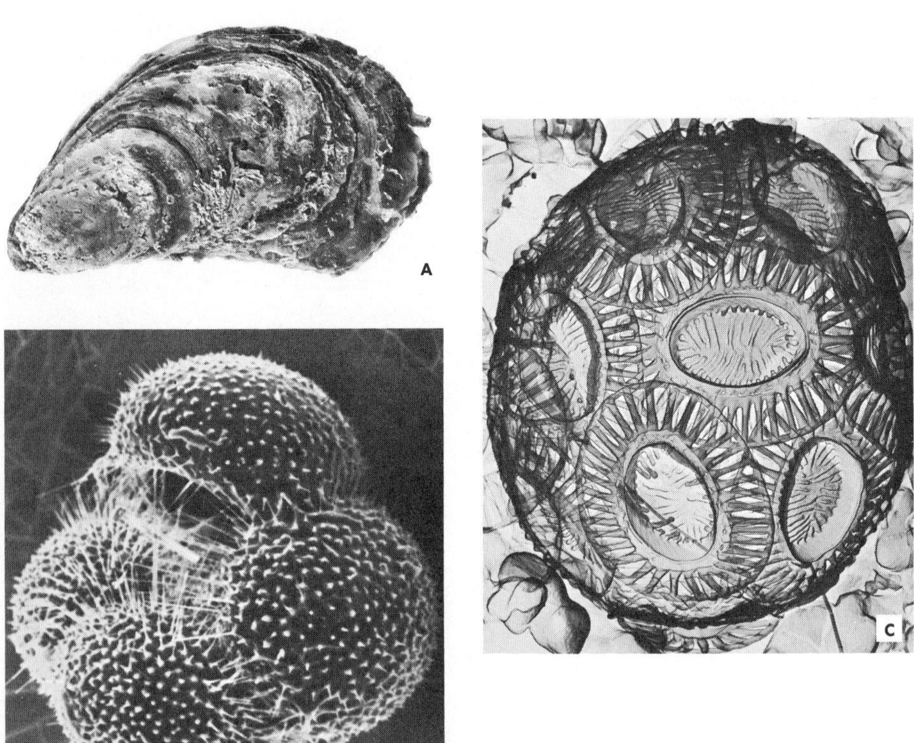

FIG. 5-5 High-magnesium calcite-depositing marine organism. The echinoderms are echinoids (sea urchins), crinoids, and starfishes. This is *Centrechinus antillarum*, a tropical Atlantic species of purple sea urchin (diameter about 15 centimeters). High-magnesium calcite shells can contain up to 15 percent magnesium. (Courtesy Carolina Biological Supply Company.)

Dolomite, a carbonate mineral having equal numbers of calcium and magnesium atoms in its crystal lattice, is not a primary deposit. It occurs commonly in the geologic record, however, and for this reason has aroused a great deal of investigation and speculation regarding its origin. Recently it has been discovered that in part, it is the probable reaction product of brines, formed by evaporation, with aragonite crystals deposited by organisms. This origin is compatible with many occurrences in the geologic record.

The Resistate Minerals

The sand fraction of sediments, aside from the calcium carbonate component, is composed of minerals generally resistant to weathering. The most common of these is quartz. In certain parts of the world other minerals, more resistant and more dense than quartz, are also concentrated as sands. If the concentration is sufficiently great and the minerals are of economic interest a fortune can be found on the seashore—for instance, the tin-bearing cassiterite sands and nearby ruby-bearing sands of Southeast Asia, the diamond-bearing sands of Southwest Africa, and the sands of South Carolina rich in rare-earth minerals.

Other "heavy minerals," less spectacular than those cited above, are also found in the coarse-grained fraction and have been useful in determining sources of sediments and paths of transport in the sea.

PARTICLES IN FLUIDS

Dust storms and muddy rivers are ample proof that fluids can transport particles. In this section we will review the general laws that govern the transport and deposition of particles in fluids in order to understand the laws that control marine sedimentation.

Stokes' Law

A fundamental consequence of Newton's laws of motion is that, regardless of mass or dimensions, *in a vacuum* all bodies have the same acceleration at the surface of the Earth. It is also a fundamental observation of parachutists that free fall does not result in the continuously increasing velocity predicted by a constant acceleration, but that a constant *terminal velocity* is reached, at which point air drag prevents further increase in velocity and acceleration becomes zero. This terminal velocity is a function of the density and shape of the body and the density of air, which varies with temperature and height.

The fall of very small sand- and clay-size particles through air and water is controlled by the same law, and for such small sizes this rule is formalized as *Stokes' Law:* For spherical particles of varying size but of the same density falling through a fluid, the fall velocity of each particle is proportional to the square of its radius (see Table 5-2). Hence, larger particles will settle out of a given column of water or air faster than smaller particles. Corrections must be made, however, if the particles have a variety of shapes, since the drag will be greater for flat, platy particles than for spherical ones. Also, in water, mineral particles having a diameter greater than 25 microns obey more complicated laws as a result of increased resistance of the fluid to the particles due to impact with the water as they descend.

Table 5-2 Stokes' Law of Settling Velocities

$$\text{Stokes' Law: } v = \frac{D^2(\rho_p - \rho_f)g}{18\eta}$$

v = velocity of fall (cm/sec)
D = diameter (cm)
ρ_p = density of particle (g/cm^3)
ρ_f = density of fluid (g/cm^3)
g = acceleration of gravity (cm/sec^2) = 980 cm/sec^2
η = viscosity (poise = g/sec cm)

For spheres of quartz (ρ_p = 2.6 g/cm^3) falling through ocean water (ρ_f = 1.0 g/cm^3) at a temperature of 10°C (η = 0.0140 poise). Stokes' Law holds for particles up to 100 microns radius (200 microns diameter). Above that size a modification (Oseen's equation) is used to account for nonlaminar behavior of the fluid due to impact by the particle.

Radius (microns)	Velocity (cm/sec)	Time To Fall through 4,000 Meters
1	0.00025	51 years
10	0.025	185 days
100 (STOKES')	2.5	1.8 days
100 (MODIFIED OSEEN'S)	1.7	2.7 days

Motion in Fluids

There are two main types of motion in fluids: laminar flow and turbulent flow. *Laminar* flow, as the name implies, involves the coherent movement of water molecules in stream lines. Particles suspended in the fluid move with the velocity of the laminar layers while also settling out under the force of gravity according to Stokes' Law. In *turbulent* flow the molecules of the fluid and the associated particles do not move along stream lines. Since there is turbulence in large fluid bodies like the oceans and the atmosphere, what then is the role of turbulence in sediment transport? Turbulence of itself does not keep particles in suspension more efficiently than a fluid with either laminar or negligible motion, *if* a continuous source of particles is not available (except where the moving parcel of fluid is bounded by fluid with markedly different properties). If, however, a source of particles remains available, then the net effect is to approach a steady state of quantity and size distribution in the fluid; this state will be characteristic of the rate of movement of the fluid. Empirical relationships between the fate of various sizes of particles and the velocity of flowing water have been observed (Fig. 5-6).

FIG. 5-6 Estimated current velocities for erosion, transportation, and sedimentation of different-sized particles (diameter in millimeters). Cohesive materials are typically muds composed of clay minerals and organic-rich sediments. 100 centimeters per second = 3.6 kilometers per hour = about 2 miles per hour. (After Heezen and Hollister, 1964, based on F. Hjulström, 1955.)

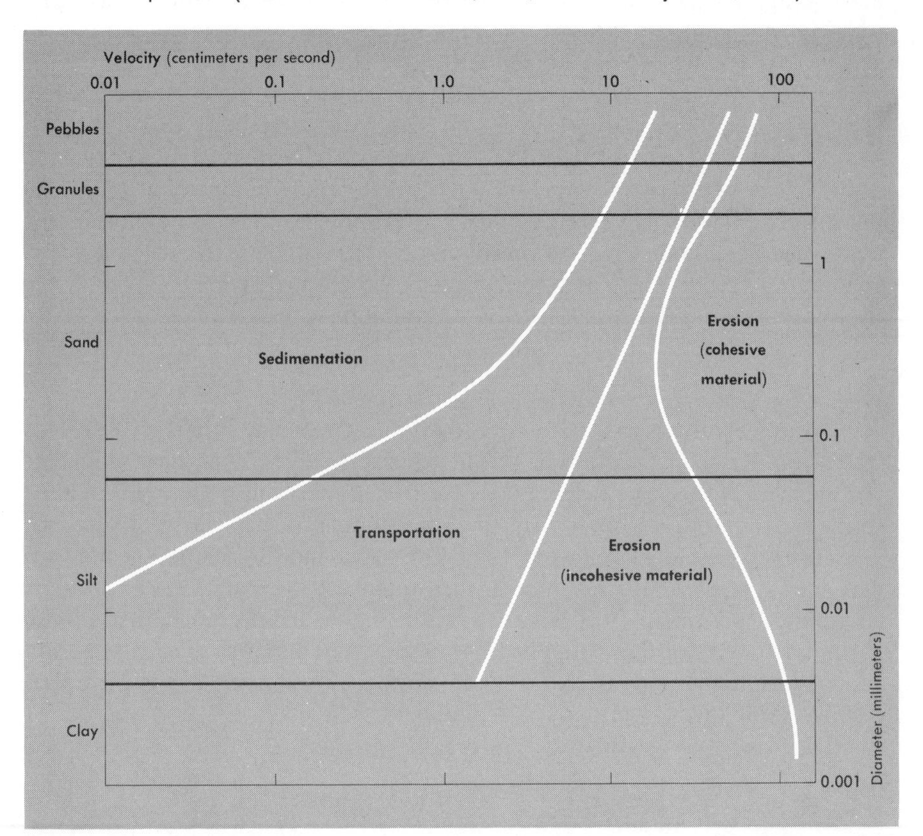

TRANSPORT OF PARTICLES TO THE OCEANS OTHER THAN BY STREAMS

The major sources of sediments to the oceans are the atmosphere, glaciers, and streams. The last of these is by far the most important as discussed in the beginning of this chapter, but it is obvious that the contributions from the other sources, where recognizable, provide valuable information on climate and wind patterns.

Atmospheric Transport

As with flowing water, an air mass moving across land will develop, as a function of velocity, a steady-state concentration of suspended material of various sizes. Once the air mass is over water, the particles should be expected to fall out in the ocean, following Stokes' Law. The dust-laden air parcels, however, are bounded above and below by air of such different properties that turbulence within the dusty air parcel keeps the fine particles suspended for greater distances than expected by simple gravity settling. Figure 5-7 shows the dust-storm patterns observed at sea, and in particular shows the strong extension from the Sahara Desert over the Atlantic Ocean.

Particles smaller than about 15 microns (0.015 millimeters) will remain aloft so long, according to the settling velocity predicted by Stokes' Law, that the main mechanism of removal from the atmosphere is entrapment in rain and snow. That is why radioactive fallout injected into the lower portion of the atmosphere, called the troposphere, is removed, for any latitude, more efficiently in areas of high precipitation and less efficiently over deserts.

If particles are injected into the stratosphere (above about 12 kilometers), as in violent volcanic explosions such as Krakatoa near Indonesia in 1883 or in the detonation of highly energetic nuclear devices, they must settle through the stratosphere without help of precipitation. Hence, these particles are distributed worldwide. Once in the troposphere, however, they are most effectively transported to the Earth's surface by precipitation.

Glacial Transport

Continental glaciers, as in Antarctica and Greenland, are efficient agents of erosion and sediment transport. Where glaciers end on land, their sediments become part of the general supply of debris to streams and thus may be diluted to the point that they cannot be distinguished in marine deposits. But when the glaciers end at the edge of the sea, as in Antarctica and Greenland, the debris is deposited directly on the sea floor. If meltwaters are the main form of terminal transport, deposition in fjords and deltas is common, and the fate of the particles is similar to that of those supplied by streams. Grains of sand, however, will show pressure-induced marks acquired during abrasion by the glacier, thus distinguishing them from purely stream-borne sediments.

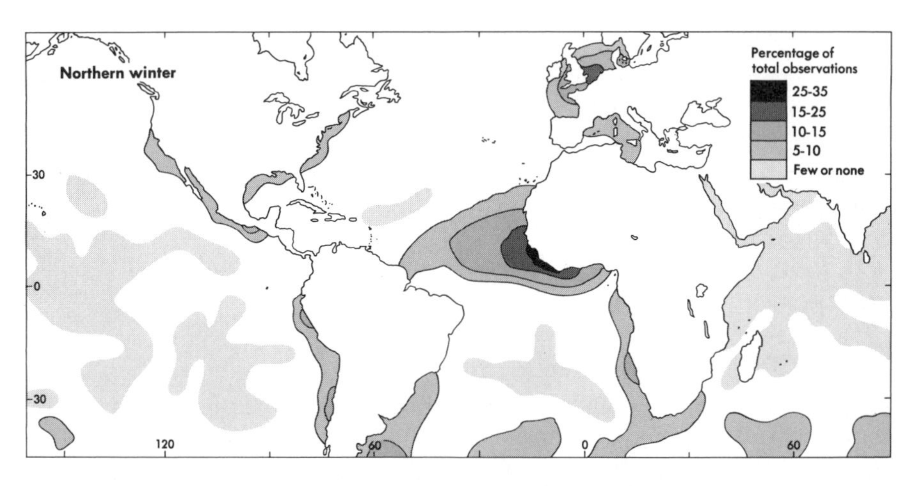

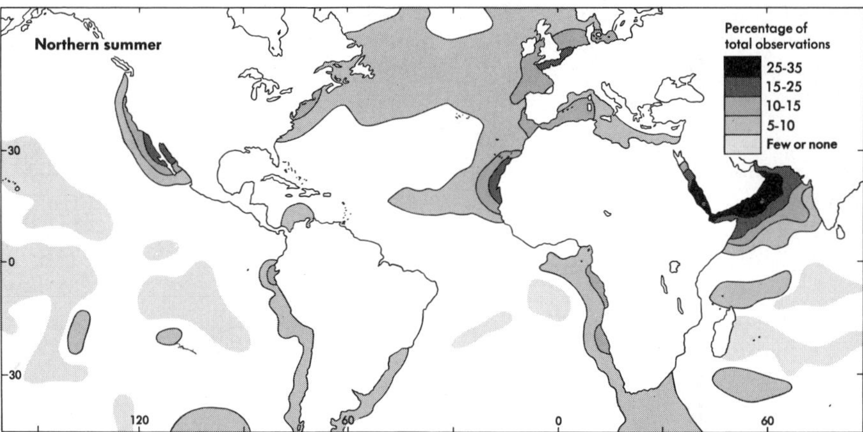

FIG. 5-7 Frequency of haze over the ocean produced by wind-borne dust from the continents. There is evidence that some dust from Africa travels as far west as Barbados and beyond. (After Arrhenius, 1963.)

Where the ends of the glaciers break off into the sea (a process called *calving*), icebergs are formed; since these may move far out to sea before melting completely, deposits of coarse material can occur in the deep ocean. Such glacial marine deposits fringe Antarctica and Greenland, and boulders have been found in deep-sea sediments as far as icebergs drift toward lower latitudes.

TRANSPORT OF PARTICLES WITHIN THE OCEANS

Once particles are brought to the oceans they may be redistributed by a variety of forces acting in the oceans. These may strongly modify the initial distribution patterns controlled by the source of supply from the continents.

Surface Currents

The prevailing winds acting on the ocean surface set up the major surface-current patterns. These are shown for the Atlantic Ocean in Fig. 3-5, Chapter 3, the most prominent feature in the North Atlantic being the Gulf Stream. The surface currents move with velocities as great as three knots (that is, nautical miles per hour), so that complete circulation of the surface water is accomplished in a few years. The fine-grained particles are carried with these currents, and with their long settling times, they have a more generalized distribution than coarser-grained particles.

Turbidity Currents

The transport of sediment downslope, primarily along submarine canyons, occurs when a sediment slurry moves as a coherent fluid with a density greater than that of sea water. Such a fluid in motion is called a turbidity current. A slide or slump of sediment somewhere on the continental margin initiates the action. The triggering mechanism of downslope movement may be an earthquake, a hurricane hitting nearshore, or a high-sediment discharge by streams.

The movements of turbidity currents resulting from earthquakes have consequences of concern to man. For instance, the Grand Banks (near Newfoundland) earthquake of 1929 resulted in downslope movement of sediment that broke transatlantic cables in that region. Turbidity currents deriving from the slump broke additional cables downhill from the slump. Figure 5-8 shows an analysis of this event.

FIG. 5-8 (Facing page) The Grand Banks slump of 1929. In 1929 an earthquake on the Grand Banks near the Laurentian Channel resulted in the breakage of a number of trans-oceanic cables. All cables within 60 miles south of the epicenter of the earthquake broke instantly due to the initial slump. Turbidity currents resulting from the slump seem to have moved down three separate channels and broken additional cables beginning 59 minutes after the earthquake occurred. (Top) A map of the cables and lines of movement. Contours are in fathoms; one fathom = 1.83 meters. (Bottom) A profile along the dotted line on the map shows the disturbed sediments and the "sole" plane of slumping sediment along which movement occurred. (After Heezen and Drake, 1964.)

Turbidity currents undoubtedly are the main mechanism of transport of material to the abyssal plains, as is indicated by layers of silt and sand and by the presence of plant remains from nearshore environments that were swept out by the currents. Graded bedding due to the varied settling rates of fine and coarse particles, and horizontal gradients in size and thickness due to the decreasing strength of the currents away from the source are also common features of turbidity current deposits.

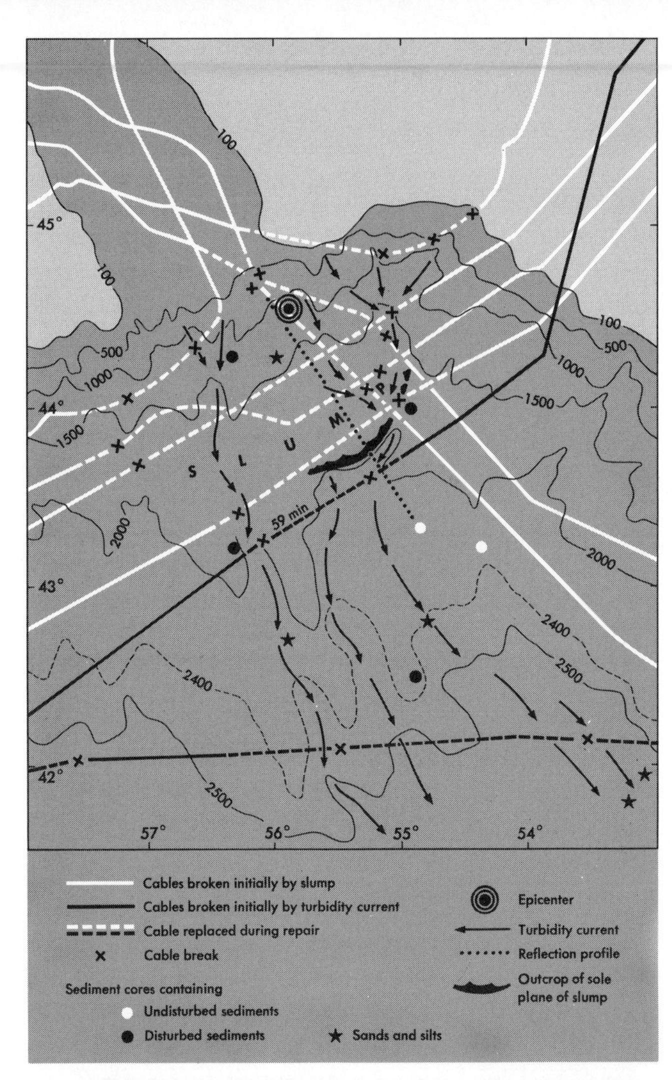
100
100
100
500
1000
1500
500
1000
1500
2000
2000
2400
2500
2400
2500
59 min
S
L
U
M
P
45°
44°
43°
42°
57°
56°
55°
54°
Cables broken initially by slump
Cables broken initially by turbidity current
Cable replaced during repair
Cable break
Sediment cores containing
Undisturbed sediments
Disturbed sediments
Sands and silts
Epicenter
Turbidity current
Reflection profile
Outcrop of sole plane of slump

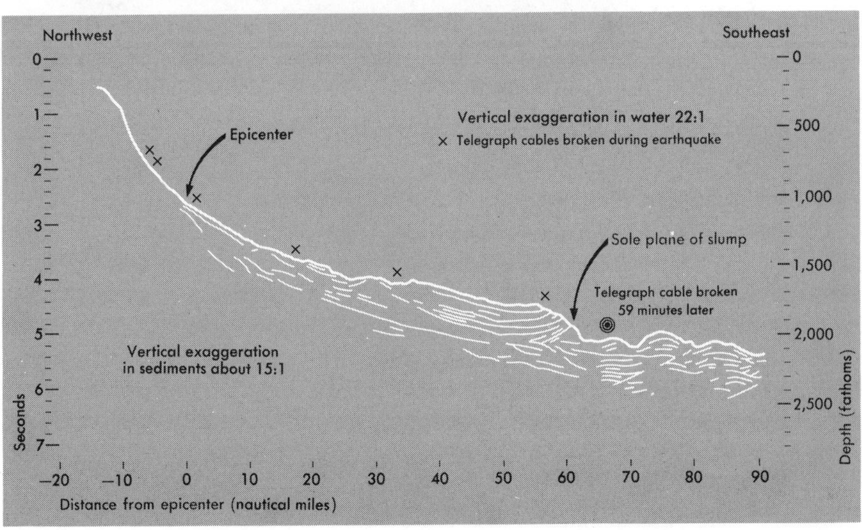
Northwest
Southeast
Vertical exaggeration in water 22:1
× Telegraph cables broken during earthquake
Epicenter
Sole plane of slump
Telegraph cable broken 59 minutes later
Vertical exaggeration in sediments about 15:1
Seconds
Depth (fathoms)
Distance from epicenter (nautical miles)
0
1
2
3
4
5
6
7
0
500
1,000
1,500
2,000
2,500
−20
−10
0
10
20
30
40
50
60
70
80
90

Bottom Currents

Oceanographic observations of various kinds indicate the presence of currents on the bottom of some parts of the ocean that are strong enough to move sediments. They are five main types of evidence: (1) In order to maintain the balance of water moving in the Atlantic Ocean Basin, strong northward-moving bottom currents must be postulated for the western South Atlantic to balance the shallower southward transport. (2) Floats that can maintain their location at a designated depth in the ocean show strong currents even at great depths, and current meters have been used actually to measure currents at the bottom. These indicate velocities up to several knots. (3) Photographs of the bottom show features that can be attributed only to strong currents. These are ripple marks (Fig. 5-9), scour marks, linear features formed in the direction of current movements, bare rock, and coarse residues due to winnowing of fines. (4) Some sediment cores raised from the deep ocean floor show strong cross-stratification, indicating the action of currents. (5) The movement of sediment can be seen by the presence of suspended material in the deep oceans. It is possible to determine the concentration of fine particles in sea water by observing the scattering of light, or by actually recovering the material by passing the water through filters with small pores, or by centrifuging it. Studies have been made of various marine environments by all these methods. Fine material from the Po River, for instance, has been traced as a deep layer in the Adriatic Sea. More recently, by means of measurement devices for scattered light, cloudy, or "nepheloid," layers have been observed that are associated with strong bottom currents over thick sediment piles.

FIG. 5-9 Ripple marks with wave lengths of 20 to 30 centimeters in two directions, indicating deep currents with variable directions. Foraminiferan tests are found in the troughs of the ripples. This photograph was taken at 2,333 fathoms (4,250 meters) in the Drake Passage (57°28′S. 64°51′W) southeast of Cape Horn. (Courtesy H. Grant Goodell.)

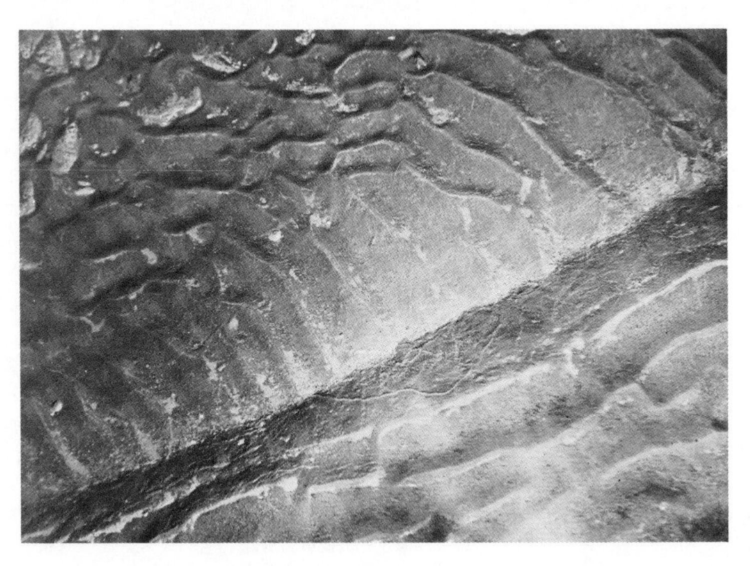

deep-sea deposits

The deep ocean is the repository of materials derived from a number of sources and transported through a number of different pathways. Sediments accumulate slowly in the deep ocean and carry with them a record of contemporaneous conditions in the oceans and on the continents. The proper interpretation of these deposits tells us a great deal about Earth history: climatic variations, patterns of ancient oceanic and wind currents, the denudation of the continents, and structural history.

CLASSIFICATION OF DEEP-SEA DEPOSITS

The first detailed study of the deposits at the bottom of the deep ocean was made at the end of the nineteenth century by Sir John Murray and A. F. Renard. Using samples that had been obtained from all the world oceans by the voyage of H. M. S. *Challenger* in the early 1870's, they classified the deep-sea sediments in terms of the dominant components obvious to the unaided eye. Hence the terms "red clay," "globigerina ooze," and "siliceous ooze" were coined. These terms remain valuable for the most general descriptions, although we have now become aware of the complexity of the mineralogy of deep-sea sediments. For instance, the term "red clay," as originally used, not only subsumed a wide range of clay minerals but was also applied to sediments primarily composed of other nonclay silicate minerals. "Globigerina ooze"

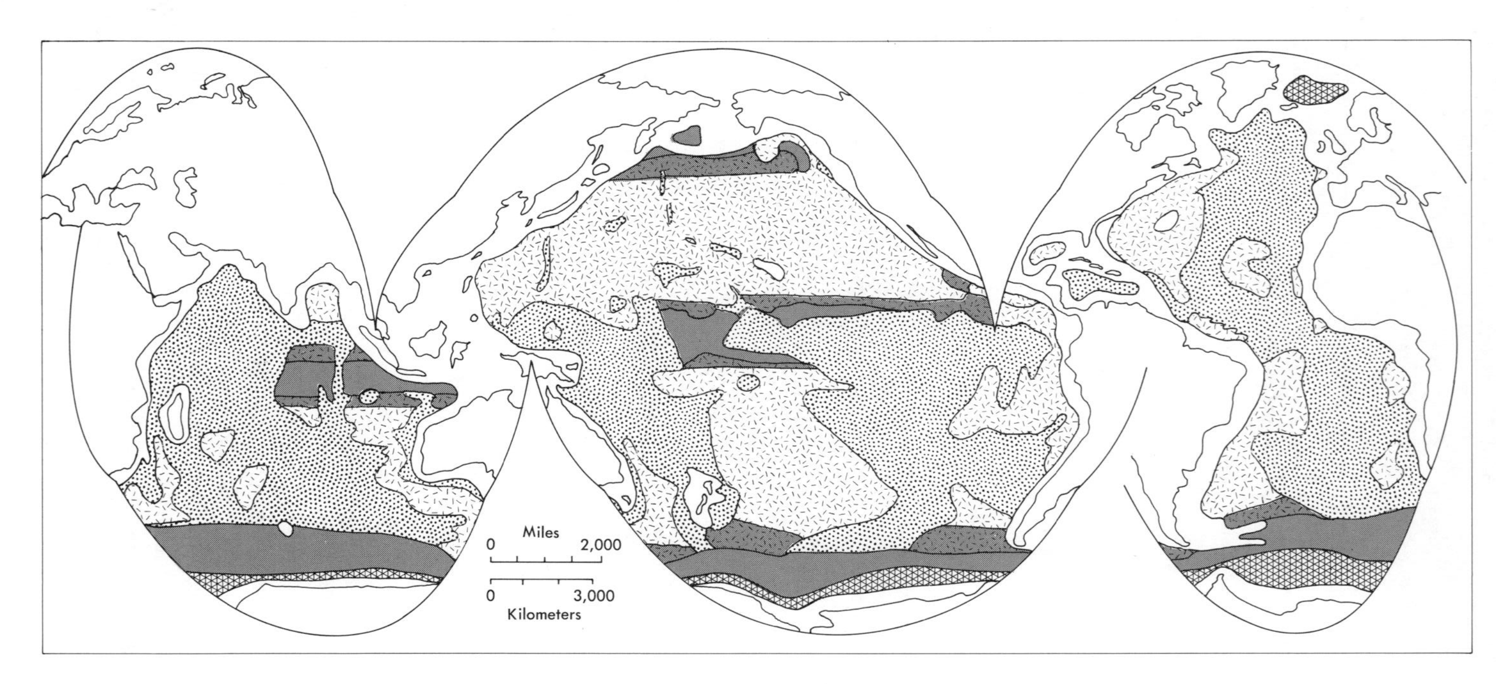

FIG. 6-1 Distribution of the major components in Recent deep-sea sediments. Calcareous ooze represents areas having greater than 30 percent calcium carbonate and siliceous ooze represents areas having greater than 30 percent siliceous fossils. (Adapted from W. Berger and F. Shepherd.)

included almost all sediments rich in calcium carbonate that were composed of coccoliths and the tests of foraminifera. Siliceous oozes were generally of two types, those rich in radiolarian tests and those rich in diatom tests.

In addition to the soft sediments the *Challenger* also recovered rounded brown-black, dusty-appearing nodules, which were determined to be composed primarily of manganese and iron oxides. Rock boulders were also caught in the dredges indicating both submarine volcanic activity and long-range transport of continental debris.

The general distributions of the major sedimentary components are shown in Fig. 6-1. A closer look at the makeup of deep-sea deposits and the relationship of their distribution to properties of the ocean and its margins, some of which have been discussed in Chapter 5, is the goal of this chapter. Table 6-1 provides a framework within which the deep-sea deposits can be discussed.

Table 6-1 Classification of Deep-sea Deposits

1. *Pelagic biogenic:* The remains of organisms: calcareous and siliceous tests and organic and phosphatic material.

2. *Nonbiogenic:* The sedimentary components not originating from life processes in the ocean.

(a) *Pelagic detrital:* This term is restricted to nonbiogenic components originating near the surface of the ocean and deposited on the sea floor by settling.

(b) *Bottom-transported detrital:* This material, typically transported by turbidity and bottom currents, contains silts and sands and remains of shallow-water organisms and land materials. There is also a fine-grained component that may not be easily distinguishable from the pelagic detrital component.

(c) *Indigenous deposits:* These are derived within the ocean basin itself by a variety of processes, such as submarine reaction of volcanic material with sea water, migration and reconstitution of materials in the sediments, and weathering of volcanic materials exposed above sea level.

CALCIUM CARBONATE

Foraminiferan tests, coccoliths, and pteropod tests are the principle constituents of the calcium carbonate ($CaCO_3$) in deep-sea sediments. The foraminiferan shells (Fig. 5-6B) found in deep-sea sediments are all composed of calcite. Coccoliths (Fig. 5-6C), the calcite tests of the algal family Coccolithophoridae, account for most of the fine-grained (less than 30 microns in diameter) carbonate material in deep-sea sediments. Pteropods are mollusks with aragonitic shells (Fig. 6-2). Their remains are preserved in deep-sea sediments at water depths of generally less than 3,500 meters. The term "globigerina ooze" applies to deep-sea sediments that are rich in a mixture of foraminiferan tests and coccoliths. A wide range of species of each group are found.

FIG. 6-2 (Above and left) Pteropod tests from the South Atlantic. Pteropods are mollusks that deposit aragonitic shells. These shells are easily seen in deep-sea cores because of their relatively large size (about 1 to 2 millimeters). They are not generally found in sediments raised from water depths greater than 3,500 meters. (Specimens courtesy Dr. C. Chen.)

FIG. 6-3 (Below) The distribution of calcium carbonate on the deep ocean floor of the Atlantic Ocean. The highest values are associated with the ridges and areas of high biological productivity such as the Cape Basin and along the Gulf Stream. The high concentrations around Bermuda and Bahamas are due to shallow-water deposits. (After Turekian, 1965.)

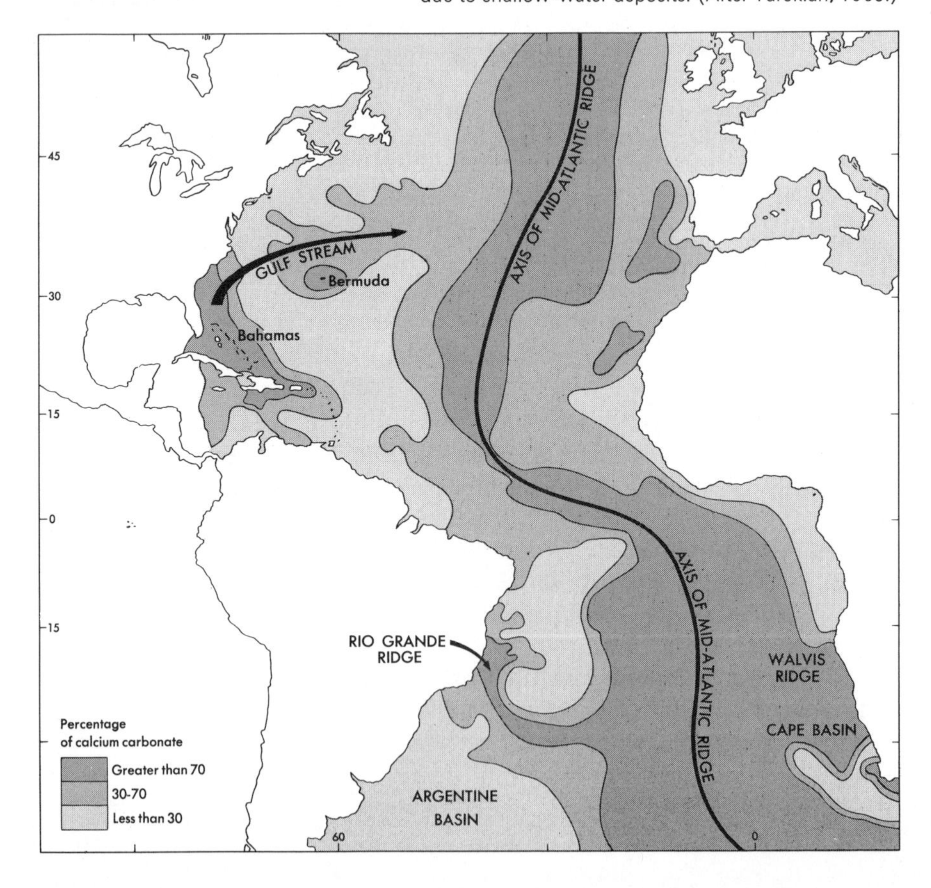

Distribution in Deep-Sea Sediments

The calcium carbonate concentration of deep-sea sediments is a direct reflection of the relative rates of accumulation of clay and calcium carbonate. The regional distribution of calcium carbonate concentration in deep-sea sediments shows three major features, as seen in the Atlantic Ocean (Fig. 6-3): (1) The topographic high points have sediments that are generally rich in calcium carbonate. (2) In some, but not all, areas where the ocean surface is biologically very productive, high calcium carbonate concentrations in the bottom sediments occur even at great depth. This is the case in the sediments along the eastward path of the Gulf Stream at about 40 degrees north latitude and beneath the area of oceanic upwelling off the coast of southwest Africa. (3) In contrast, the Argentine Basin shows a virtual absence of calcium carbonate in the bottom sediments at any depth, despite the fact that a normal biological productivity occurs in the surface waters.

This pattern is repeated in the East Pacific Ocean (Fig. 6-4). Here the sediments of the East Pacific Rise are high in calcium carbonate as are the sediments under the highly productive waters along the East Pacific Equatorial current. But in the North Pacific Ocean, the sediments are low in calcium

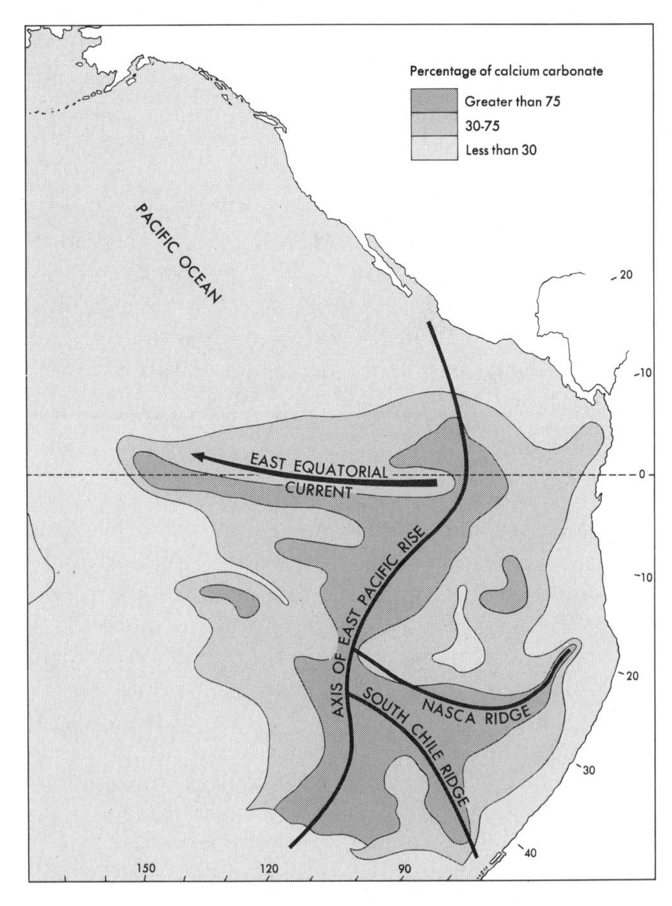

FIG. 6-4 The distribution of calcium carbonate on the deep ocean floor of the east Pacific Ocean. Similarly to the Atlantic Ocean, the highest values are associated with ridges and areas of high biological productivity such as the east equatorial Pacific. (After Bramlette, 1961.)

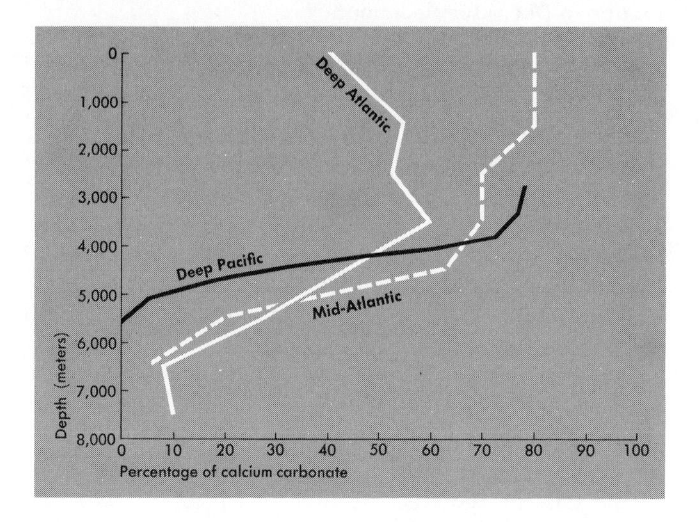

FIG. 6-5 The distribution of calcium carbonate in the tops of deep-sea sediment cores as a function of depth of water. The curves connect average values for 1,000-meter intervals for the Atlantic and 500-meter intervals for the Pacific. The "compensation depth" is the approximate depth at which the restricted accumulation of calcium carbonate results in a sharp decreasing trend in calcium carbonate concentration with depth. It is deeper in the mid-Atlantic region than in the deep Pacific. The form of the curve for all data in the deep Atlantic reflects in part the large variations in the rates of clay accumulation as well as the simple depth effect on calcium carbonate accumulation. (After Bramlette, 1961, and Turekian, 1965.)

carbonate, as in the Argentine Basin. In general, however, with increasing depth there is a gradual decrease in calcium carbonate until about 4,500 meters. At depths greater than this, the concentration drops to extremely low average values (Fig. 6-5). The point at which the calcium carbonate concentration of deep-sea sediments decreases rapidly with depth is commonly called the "compensation depth." At this depth the rate of supply of calcium carbonate is ideally just compensated by an equal rate of solution.

Why the "Compensation Depth"?

The strong water depth control on calcium carbonate concentration in deep-sea sediments has intrigued scientists since its discovery. The most attractive explanation at one time for most but not all of the observations was based on the level at which supersaturation relative to calcium carbonate gave way to undersaturation with respect to this component. This idea was amenable to laboratory testing and a surprising result was obtained. At depths greater than 500 meters, to judge from these laboratory experiments, sea water is undersaturated with respect to calcium carbonate generally and thus it should dissolve. Hence, if the compensation depth were purely the point at which calcium carbonate ought to disappear into solution, the critical level would be 500 meters in many places, not 4,500 meters.

We are thus left with a kinetic model for the explanation of the dissolution of calcium carbonate at depth. Some of the first evidence for this resulted from an experiment in the Pacific in which accurately weighed polished spheres of calcite were precisely located along a wire suspended in the ocean. When the calcite spheres were recovered after being subjected to the dissolving action of sea water for a while, it was found that the spheres below 3,800 meters showed more dissolution than the spheres above this depth (Fig. 6-6). Clearly the rates of solution were related to the changing properties of the water column, even though it was all undersaturated (below 500 meters) with respect to calcite. The experiment was repeated with actual foraminiferan tests, and the same effects were observed. The foraminiferan tests in deep-sea deposits show a distribution that contributes to the ideas about the behavior of calcium carbonate in ocean-water profiles. Two major types of foraminiferan tests can be distinguished on the basis of the properties of shell structure: (1) the fragile spinose tests, and (2) the dense relatively flattened nonspinose tests.

Above a certain depth (about 3,800 meters in the central Pacific) both types exist in the sediments in large quantities. Below that depth virtually all of the fragile foraminiferan tests are dissolved, leaving behind a residuum of only the robust tests. This boundary is called the *lysocline* and clearly represents a change in efficiency of dissolution with depth in the ocean. It has been determined in laboratory experiments that the lysocline can indeed be explained by a sharp change in the degree of undersaturation in the ocean.

We can now explain the compensation depth according to this chemical kinetic model. Above the lysocline the deposition rate of calcium carbonate exceeds the solution rate determined by the water column properties. Once

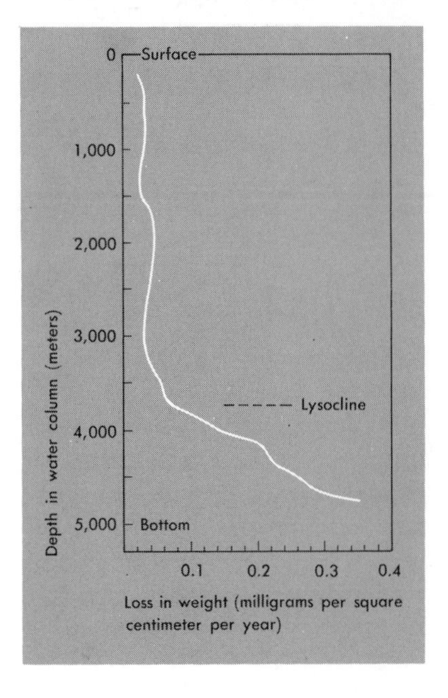

FIG. 6-6 The dissolution of calcite spheres suspended at different depths in the Pacific Ocean, 1,100 kilometers southwest of Honolulu, Hawaii (18°49′N, 168°31′W). Weight loss rate is finite but low between 500 meters and 3,800 meters. The marked increase in dissolution rate occurs below 3,800 meters. This has been called the lysocline. (After Peterson, 1966.)

calcium carbonate begins to accumulate, the local pore waters and the micro-boundary layer at the sediment-water interface become saturated with respect to calcium carbonate. This effectively stops the dissolution of falling tests as they reach the bottom.

At the lysocline, however, the solution rate is close to the deposition rate and calcium carbonate accumulation is diminished. If the solution rate equals the deposition rate, the compensation depth and the lysocline are identical.

In areas of high surface organic productivity it is possible for the deposition rate to exceed the solution rate at all depths sampled below the lysocline. Under these conditions the establishment of a saturated pore water and microlayer at the sediment-water interface will again act to preserve the calcareous tests as they rain on the ocean floor, and the compensation depth will be deeper.

OTHER BIOGENIC DEPOSITS

Silica

Four groups of organisms deposit siliceous tests found in deep-sea sediments: diatoms, radiolarians, sponges, and silicoflagellates. The most important of these in terms of quantity are diatoms (Fig. 6-7) and radiolaria (Fig. 6-8). Diatoms, like the coccolithophorids, are photosynthetic organisms and radiolarians, like the foraminifera, are particle eaters.

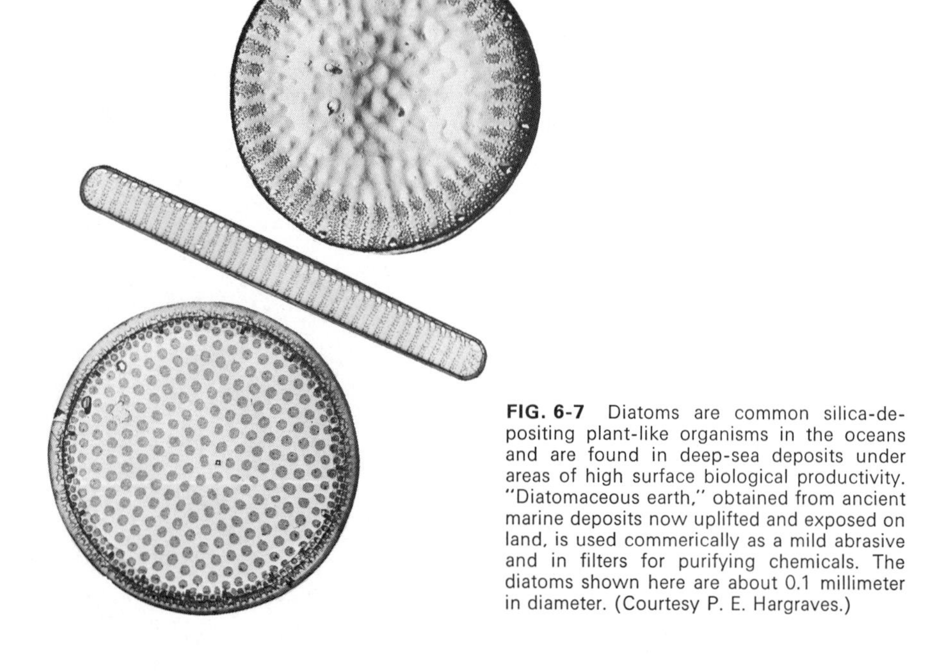

FIG. 6-7 Diatoms are common silica-depositing plant-like organisms in the oceans and are found in deep-sea deposits under areas of high surface biological productivity. "Diatomaceous earth," obtained from ancient marine deposits now uplifted and exposed on land, is used commerically as a mild abrasive and in filters for purifying chemicals. The diatoms shown here are about 0.1 millimeter in diameter. (Courtesy P. E. Hargraves.)

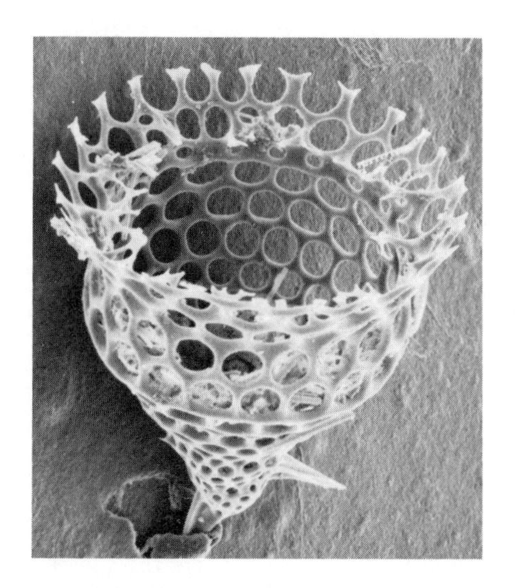
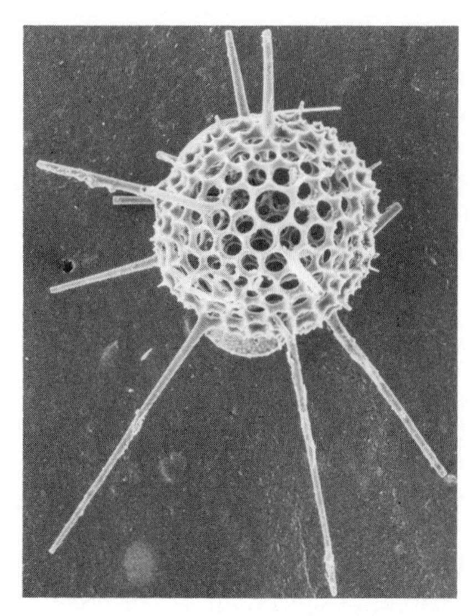

FIG. 6-8 Radiolarians from deep-sea sediments around Antarctica. Radiolarians deposit siliceous tests but are not photosynthetic organisms like diatoms. Length about 0.1 millimeters.

Since silica, in the form deposited by these organisms, is highly soluble in sea water, the accumulation of tests on the ocean floor depends on how easily a test is dissolved on its descent through the ocean. There is no "lysocline" for silica as defined for calcium carbonate. The rate of solution depends on the size of the test, the concentration of silica in the water column, and the protective chemical bondings with organic compounds or elements such as magnesium on the test surface.

Siliceous sediments are found mainly at high latitudes, in the equatorial Pacific, and generally in regions of upwelling waters. Large diatom tests accumulate rapidly at high latitudes during the summer seasons when sunlight persists virtually all day and the nutrient supply is maintained by rapid convective overturn; in the equatorial Pacific the efficient growth of diatoms and radiolaria occurs because of a good supply of nutrients and silica from the upwelling of deeper phosphate- and silica-rich waters.

Phosphatic Deposits, Organic Deposits, and Strontium Sulfate

Since the bones of most vertebrates are composed mainly of the mineral apatite ($Ca_5(PO_4)_3OH$), and since the bone debris of fish and marine mammals are likely to sink, we would anticipate finding phosphatic deposits on the sea floor. Contrary to expectation, however, very little skeletal phosphate reaches the deep ocean bottom, for either the effect of pressure on increasing the solubility of apatite and other phosphate compounds or the efficiency of biological cycling prevents preservation of the bones and other phosphatic debris during their descent through the water. Fish debris, shark teeth, and ear bones of whales have been found in deep-sea sediments, but these occurrences, although interesting, are minor in terms of the total sedimentation in the deep oceans.

In areas of upwelling and related high biological productivity the supply of the phosphate-bearing tissue to the ocean bottom may be quite high together with silica, organic material, and certain metals such as uranium. The sediment in these areas is "fluffy" but there is evidence that the accumulated phosphatic material can be reorganized to form phosphorite nodules and encrustations.

Organic compounds produced by life processes likewise are less abundant in deep-sea sediments than might be predicted from the amount produced in surface waters. Most of the organic compounds are utilized as food before they reach the bottom. Despite this extraction, however, enough reaches the bottom to support a sizeable fauna of worms, serpent stars (echinoderms), holothuria, and mollusks, to name the major groups.

Organic material accumulates more readily in sediments where the overlying water is deficient in oxygen. Most of the ocean basins have plenty of dissolved oxygen that can be used for the metabolism of organic compounds by aerobic (or molecular-oxygen-using) organisms. This supply is available because there is a relatively rapid turnover of the oceans and well-aerated waters penetrate to the deepest parts of the oceans.

Some local basins, however, can become isolated from the main deep-water circulation, and there the oxygen is used up so rapidly that the level always remains extremely low. These are called stagnant basins. Although anaerobic bacteria can make use of combined (nonmolecular) oxygen to metabolize organic material, the process is considerably less efficient than the process of aerobic metabolism, and consequently more organic material accumulates in the sediments.

The Black Sea and the Cariaco Trench off Venezuela are essentially anaerobic basins of this sort and the concentration of organic material in the sediments is as high as 3 to 5 percent. The Argentine Basin, on the other hand, has as much as 5 percent organic material in the sediments even though the oxygen level is not very low. This difference can be explained if either the organic compounds present are not easily metabolized by common marine organisms or if the rate of supply is greater than the rate of usage by bottom-dwelling organisms. On the whole, however, the organic-compound concentration in deep-sea sediments is less than 1 percent.

One group of organisms, Acantharia (related to the radiolarians), deposits tests of strontium sulfate (as the mineral celestite). In many places in the ocean surface waters they are equal to the foraminifera and radiolaria in abundance, but they have never been found in deep-sea deposits. They do occur—although only sparsely—in sediments accumulating at relatively shallow depths (less than 300 meters). Clearly, like phosphatic material, most of the strontium sulfate is dissolved high in the water column. Thus an entire unique group of organisms escapes recording in deep-sea sediments. We can see from this special case how incomplete the sedimentary record can be.

NONBIOGENIC DEPOSITS

Clays and Related Minerals

In the distribution of clay minerals in deep-sea sediments we encounter the three nonbiogenic components: the pelagic, the bottom-transported, and the indigenous.

Clay sedimentation in the Atlantic Ocean is almost entirely influenced by continental sources. Figure 6-9, for instance, shows the variation of the kaolinite/chlorite ratio in the Atlantic Ocean. This ratio is particularly informative because it represents the maximal effect of climatic zones on weathering products. Kaolinite is typical of weathering in tropical climates, conditions under which chlorite is easily destroyed. Weathering in temperate and arctic environments results in the preservation or formation of chlorite. Thus, there is a sharp decrease in the kaolinite/chlorite ratio with increasing latitude.

FIG. 6-9 The kaolinite/chlorite ratio variation in the clay-size fraction of deep-sea sediments of the Atlantic Ocean. Kaolinite is formed in intensely weathered soils typical of the equatorial region. The pattern in the sediments reflects the weathering intensity on the adjacent continents. The ratio of kaolinite to chlorite is actually the ratio of intensities of X-ray diffraction peaks. (After Biscaye, 1965.)

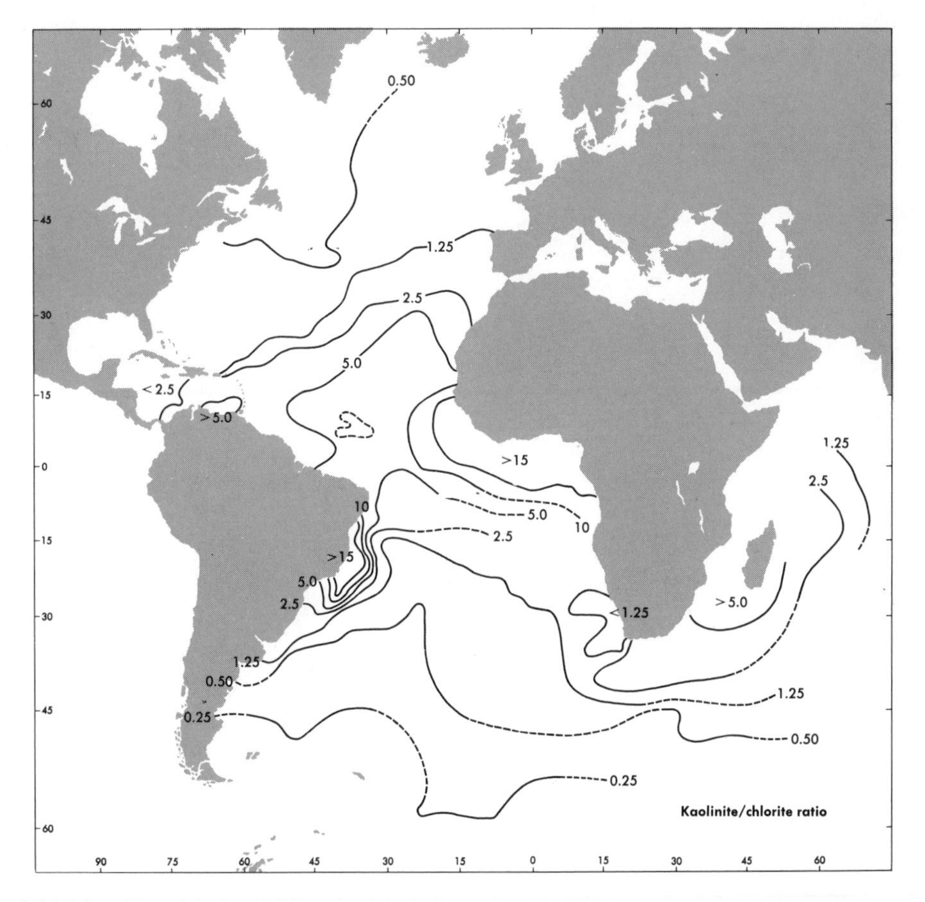

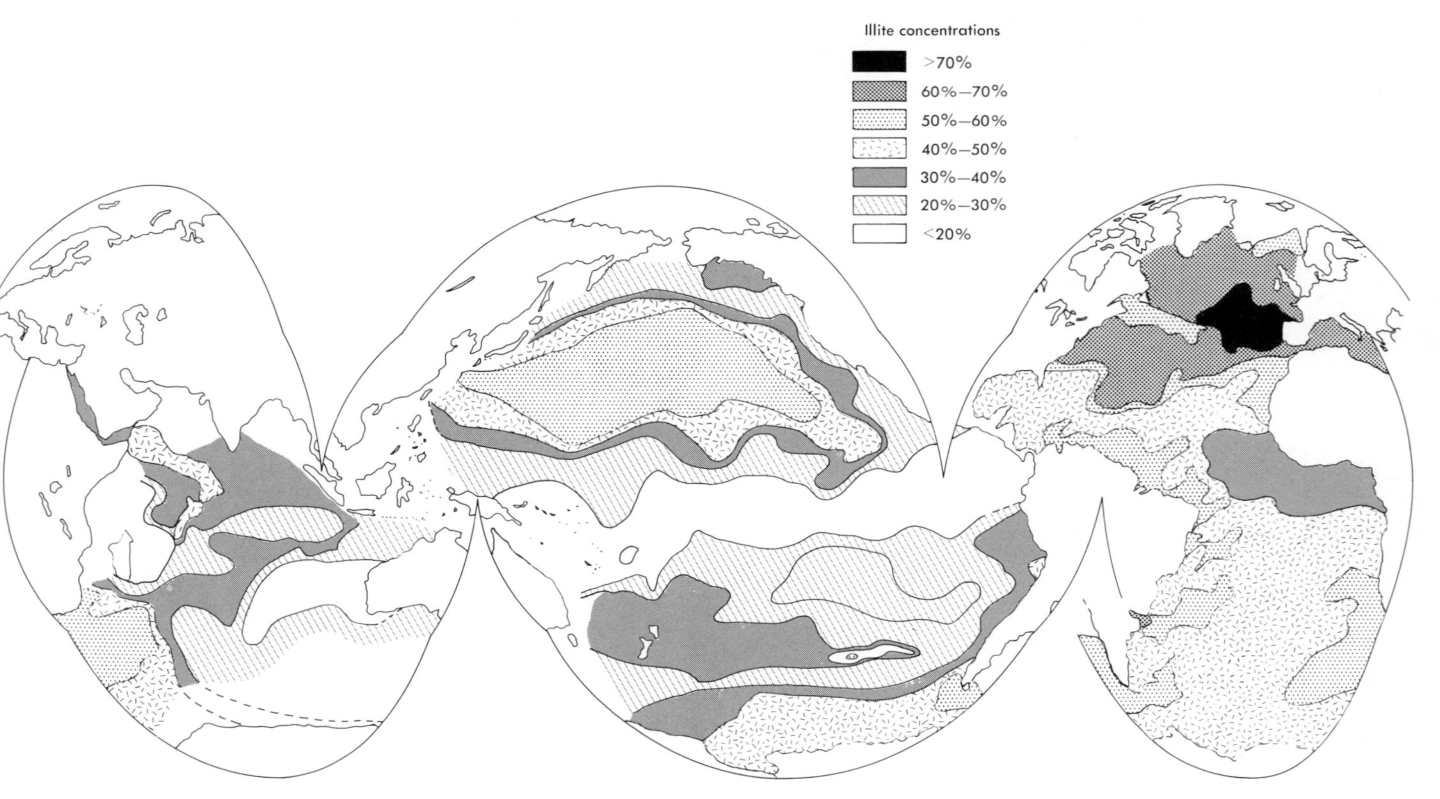

FIG. 6-10 The distribution of illite (or detrital mica) in the world oceans. (Compiled by J. J. Griffin, H. Windom, and E. D. Goldberg.)

Gibbsite, the hydrated aluminium oxide, is the common end product of intense tropical weathering, and is a principle component of the most important aluminum ore, bauxite. Its distribution in deep-sea sediments is similar to that of kaolinite, for it occurs where continent-derived materials are dominant, as in the Atlantic Ocean.

The Pacific Ocean is rimmed by deep trenches that trap bottom-transported material from the continents. This is so at the present time, but in the past, continuous canyons and other avenues of transport down the continental rise must have existed in the northeast Pacific Basin to result in the large abyssal plains in that area.

Pelagic detrital sediment (as defined in Table 6-1) thus occurs throughout most of the oceans, but its importance relative to indigenous components (as defined in Table 6-1) diminishes near the center of the Pacific where volcanic activity has been prominent. This is best seen in the relative abundance of the clay mineral illite (or detrital mica), clearly derived from land, to the other minerals of deep-sea sediments (Fig. 6-10). The most prominent of these other minerals in the Pacific is montmorillonite, which is closely associated with indigenous volcanic activity.

Volcanic Products

Three types of volcanic material exist on the ocean bottom: rock fragments, volcanic glass, and minerals (among them montmorillonite) produced by the reaction of volcanic material with hot water. Since volcanic islands are predominantly basaltic in composition, it is not surprising that by far the most common rock fragments found by dredging the ocean bottom are basalts. These samples are found mainly on the oceanic ridges and seamounts. Volcanic glass found in deep-sea deposits is also primarily basaltic, although pumice (a glass more granitic in composition) derived from volcanism on the continents is not uncommon. Basaltic glass exists either in the relatively unhydrated form called hyaloclastite or in the hydrated form, palagonite. The hyaloclastite appears to be stable for long periods of time, but in sea water the palagonite devitrifies to form both the zeolite phillipsite and montmorillonite (Fig. 6-11).

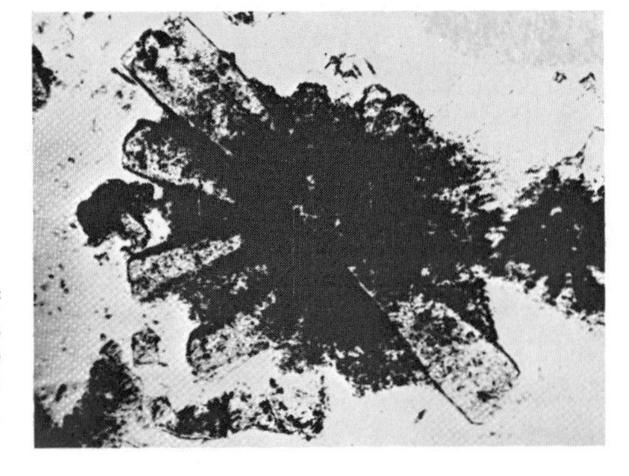

FIG. 6-11 Basaltic glass with needles of the zeolite, phillipsite, radiating from it. It is believed that phillipsite and montmorillonite are formed by the devitrification of hydrated basaltic glass. Magnification 190 times. (Courtesy E. Bonatti.)

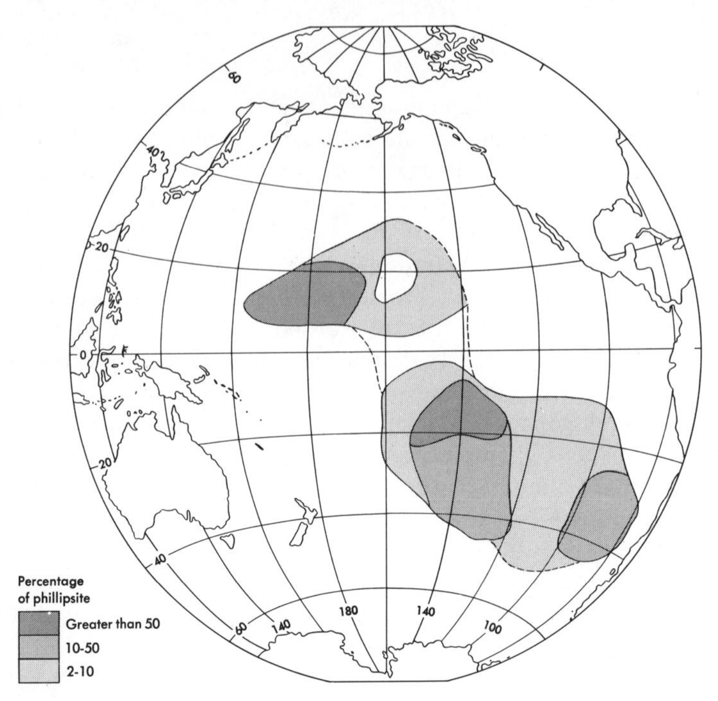

FIG. 6-12 The distribution of the zeolite, phillipsite, in the Pacific Ocean deep-sea sediments. It is clearly related to areas of high volcanic activity in the middle of the Pacific Ocean, where it can make up more than 50 percent of the silicate minerals in the sediments. The dashed lines are estimated boundaries where data do not exist. Phillipsite is very rare in the Atlantic Ocean. (After Bonatti, 1963.)

The zeolite minerals in the deep-sea sediments are presumed to be derived indigenously. The zeolites are framework-type silicates, many of which have strong ion-exchange or water and volatile adsorption capacities. They commonly occur as late-stage products of hot-water activity in volcanic areas. The most common zeolite in the Pacific Ocean sediments is potassium-rich phillipsite, which may constitute more than half of the sediment in areas with a low rate of clay accumulation and a high rate of volcanic activity. A map of phillipsite distribution in the Pacific is reproduced in Fig. 6-12.

Detrital Silts and Sands

Between the extremes of the clay-sized materials and the volcanic rock outcrops lie the silt-, sand-, and boulder-sized deposits. The transport of these materials to the ocean floor is by a variety of methods. The minerals in the silt and sand sizes include quartz, the most common mineral, and to a lesser extent, feldspars, amphiboles, and other minerals normally more subject to chemical attack during weathering than quartz.

Silts and sands are common constituents of abyssal plain deposits (Fig. 6-13). They are transported from the continental margin by turbidity currents. There is also a strong silt component transported by western boundary currents from the glacial debris from high latitudes. The minerals provided by physical

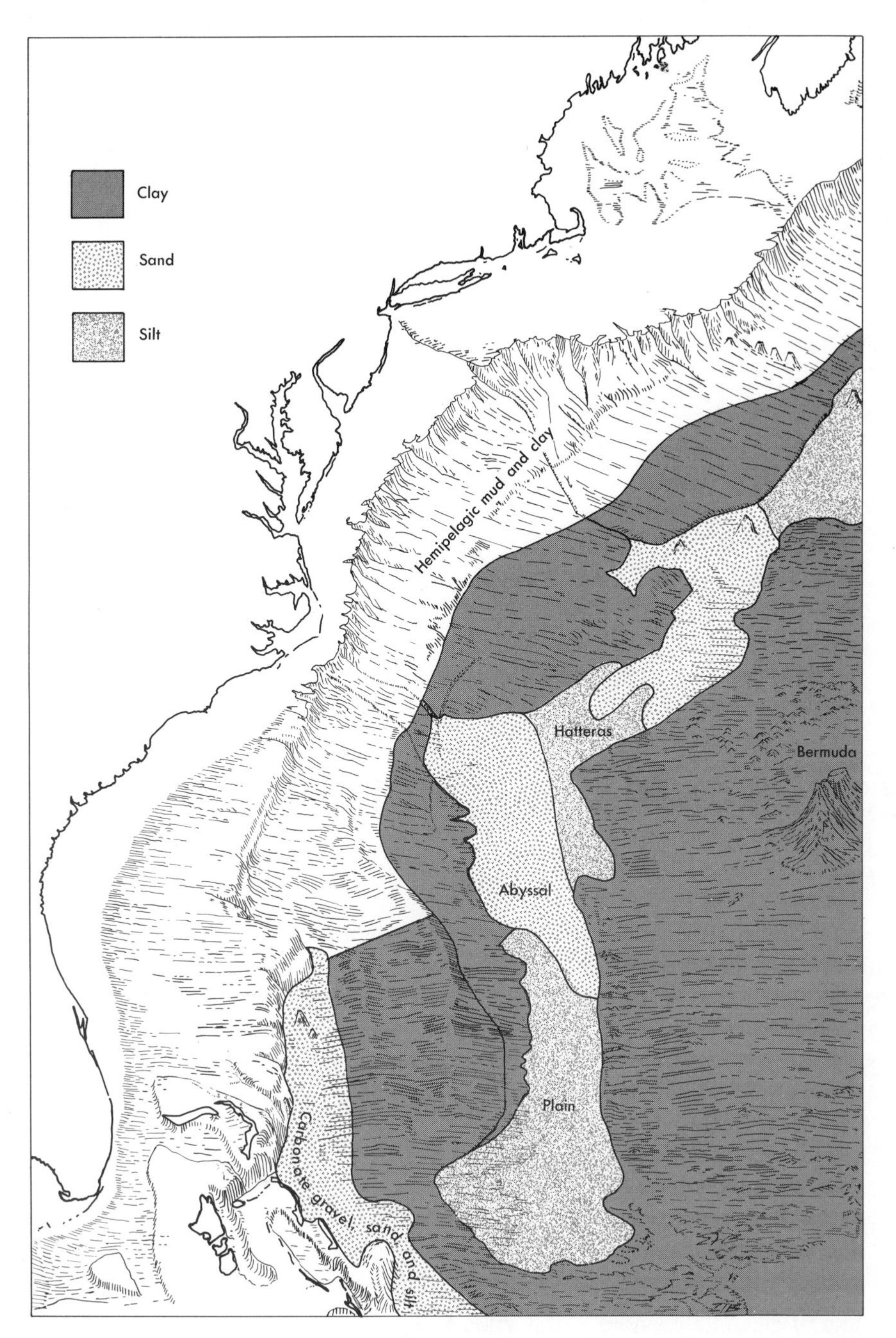

FIG. 6-13 The distribution of clay-, sand-, and silt-rich sediments in the Hatteras abyssal plain. Note the relation between major canyon systems and the distribution of the coarser-grained sediments. (After Horn, Ewing, Horn, and Delach, 1971.)

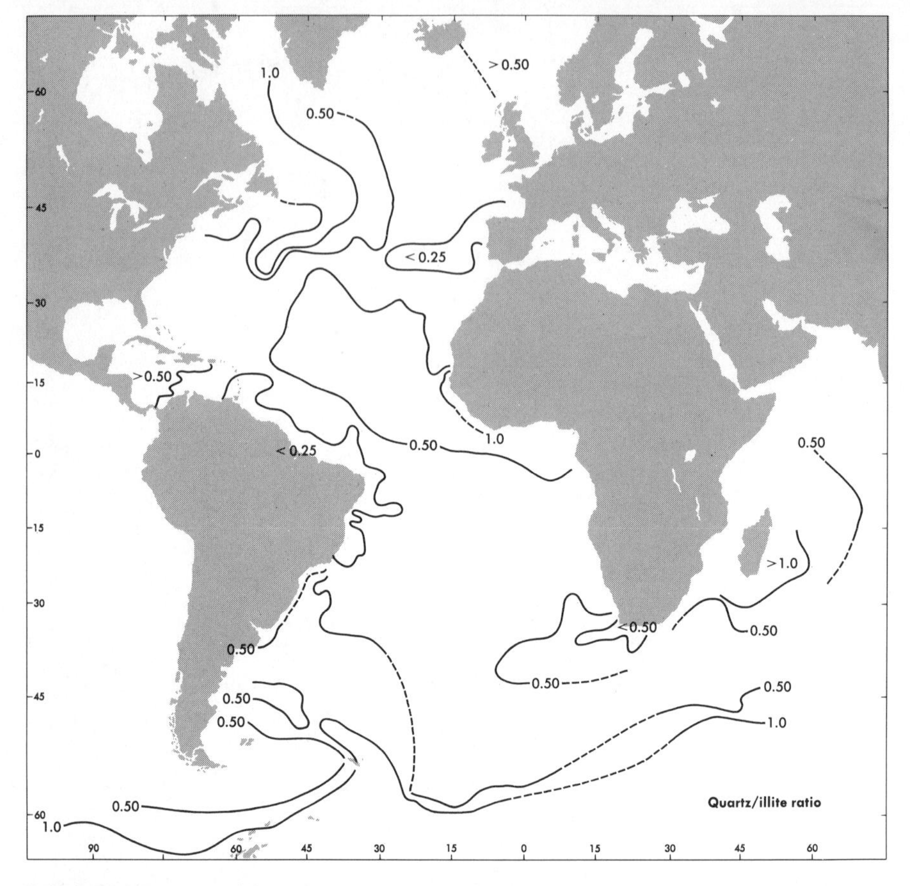

FIG. 6-14 The quartz-illite ratio variation in the fine silt-size fraction of deep-sea sediments of the Atlantic Ocean. The high quartz content off the Sahara probably represents wind-borne material (see Fig. 5-7). The high quartz at the high latitudes is due to the supply of ground quartz-bearing rock by glaciers from Greenland and Antarctica. The ratio of quartz to illite is actually the ratio of intensities of X-ray diffraction peaks. (After Biscaye, 1965.)

crushing of continental rocks by glaciers include quartz and amphibole, and their trace from these sources can be followed along the track of the western boundary currents (Fig. 6-14). A third source of detrital materials of silt size is wind transport from arid lands. Figure 6-14 shows a high level of quartz westward from the Sahara in the eastern Atlantic. Windblown quartz has also been identified in the soils of Hawaii. This must be derived from the Gobi Desert since virtually no quartz is found as part of the volcanic mineral assemblage of these basaltic islands.

METALLIFEROUS DEPOSITS

One of the tantalizing features of deep-sea deposits is the distribution of metals of possible economic importance. The most striking of these deposits are the ferromanganese nodules found throughout the world's oceans. Metal concentrations also occur in association with the major ocean ridge systems and in areas of high surface ocean biological productivity.

Manganese (or Ferromanganese) Nodules

Concretions, coatings, or nodules of hydrous manganese and iron oxides are a common feature of the deep-sea floor. The occurrence may range from coatings on minerals or coccoliths (less than 30 microns in thickness) to discrete nodules up to 850 kilograms in mass. Table 6-2 gives the average composition of manganese nodules. In the formation of nodules, manganese and iron accrete in concentric layers, sometimes mixed with foreign material such as clay, calcium carbonate, or volcanic debris. Manganese nodules have been found by dredging and by underwater photography in most parts of the oceans (Fig. 6-15). They

Table 6-2 Average Composition of Manganese Nodules

ELEMENT	WEIGHT PERCENTAGES (DRY-WEIGHT BASIS)		ELEMENT	WEIGHT PERCENTAGES (DRY-WEIGHT BASIS)	
	Pacific Ocean	Atlantic Ocean		Pacific Ocean	Atlantic Ocean
B	0.029	0.03	Fe	14.0	17.5
Na	2.6	2.3	Co	0.35	0.31
Mg	1.7	1.7	Ni	0.99	0.42
Al	2.9	3.1	Cu	0.53	0.20
Si	9.4	11.0	Sr	0.081	0.09
K	0.8	0.7	Y	0.016	0.018
Ca	1.9	2.7	Zr	0.063	0.054
Sc	0.001	0.002	Mo	0.052	0.035
Ti	0.67	0.8	Ba	0.18	0.17
V	0.054	0.07	Yb	0.0031	0.004
Cr	0.001	0.002	Pb	0.09	0.10
Mn	24.2	16.3			

After Mero, 1965.

FIG. 6-15 A section through a manganese nodule dredged from the Atlantic Ocean floor showing a core or nucleus of volcanic rock. Other materials may act as nuclei of manganese nodules including shark teeth, ear bones of whales, and even naval artillery shells. Scale at lower right equals 1 centimeter.

are, however, most common in areas with low accumulation rates of clay and calcium carbonate, specifically: (1) areas scoured by bottom currents such as submarine hills and regions like the Drake Passage between Antarctica and South America and the Blake Plateau off the southeastern United States, and (2) areas of slow pelagic clay deposition such as the central Pacific Ocean (Fig. 6-16).

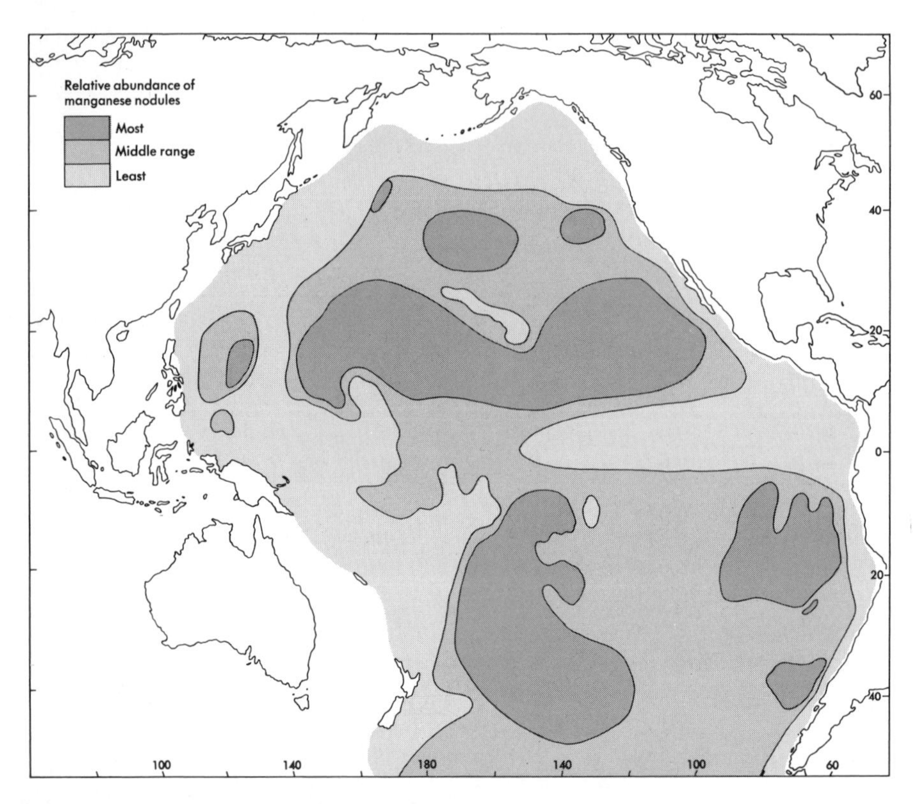

FIG. 6-16 The relative distribution of manganese nodules on the Pacific Ocean floor from the estimates of Soviet scientists. A similar general pattern may exist for the Atlantic Ocean. The ridge areas, or areas under strong surface biological productivity, where the sediments are rich in calcium carbonate, do not appear to have high concentrations of manganese nodules. (After Skornyakova and Andruchenko, 1964.)

The source of the elements for the nodules is probably twofold: (1) detrital manganese and iron oxides from the continents, and (2) manganese and iron derived from the reaction of submarine volcanic products with sea water.

The nodules and concretions grow at various rates. For instance, coatings of ferromanganese oxide can be seen on very recently deposited pteropod tests in deep-sea sediments. A twentieth-century naval artillery shell dredged from the continental margin off California had developed layers, several centimeters thick, with a high ratio of iron to manganese. Other thick ferromanganese deposits both on the continental margin and along the Mid-Atlantic Ridge have been determined by radioactive dating to have grown very rapidly. Nodules from the deep-sea floor, however, indicate growth rates slower than 5 millimeters per million years, as determined by means of radioactive dating.

Nodules appear to be lying at the interface between water and sediment but some underwater photographs indicate that sediments may cover large areas of manganese nodules. To judge from a number of successful recoveries of nodules from cores, they occur in the sediment at least as deep as one meter.

How do ferromanganese nodules and manganese-rich encrustations form? Probably the mechanisms differ among the different major types of deposits, but certain common conditions are required by the evidence of their existence. It looks as if all the manganese and iron in these oxide deposits went through a solution phase. Under the oxidizing conditions of most of the deep-sea floor, the solubility of iron and manganese is very low. These elements can be supplied in solution either by diffusion from pore waters of buried sediments, which have become reduced at depth, resulting in the increase of dissolved Fe^{+2} and Mn^{+2} (the reduced ionic species of iron and manganese), or by the flushing of hot sea water through a newly emplaced cooling basaltic deposit on the sea floor. The latter mechanism is capable of providing a large supply of iron but not of manganese because of the composition of basaltic rocks. The evidence for altered basaltic rock in deep-sea sediments has been discussed.

The migration of iron and manganese as the result of the bacterial reduction of sulfate in anoxic sediments is known to occur in nearshore deposits. It is certainly a mechanism of transport of manganese and iron from deep sediments to the oxygenated sediment-water interface where these ions will be oxidized and precipitated on the oxides. This mechanism, however, is not able to solve the problem of the high metal contents of the nodules because under the conditions described for the migration of iron and manganese, other metals will be immobilized (see Chapter 9).

The probable source for much of the manganese and iron found in the ubiquitous nodules is fine-grained particulate transport by pelagic sedimentation. Fine-grained, iron-and manganese-rich particles slowly make their way to the ocean floor. A remobilization and reprecipitation of these metals occur via the agency of specific bacteria found in association with nodules. The other metals presumably arrived at the ocean bottom in association with the highly adsorbant ferromanganese particles and were subsequently included in the growing nodules.

Ocean Ridge Related Metal Deposits

The major ocean ridge systems are characterized by higher concentrations of many metals than adjacent areas. Figure 6-17 shows the distribution of cobalt in sediments from the Atlantic Ocean. This pattern is repeated for manganese, copper, lead, barium, and several other metals. A similar pattern exists in the Pacific with concentrations of many metals higher along the East Pacific Rise than the surrounding ocean floor.

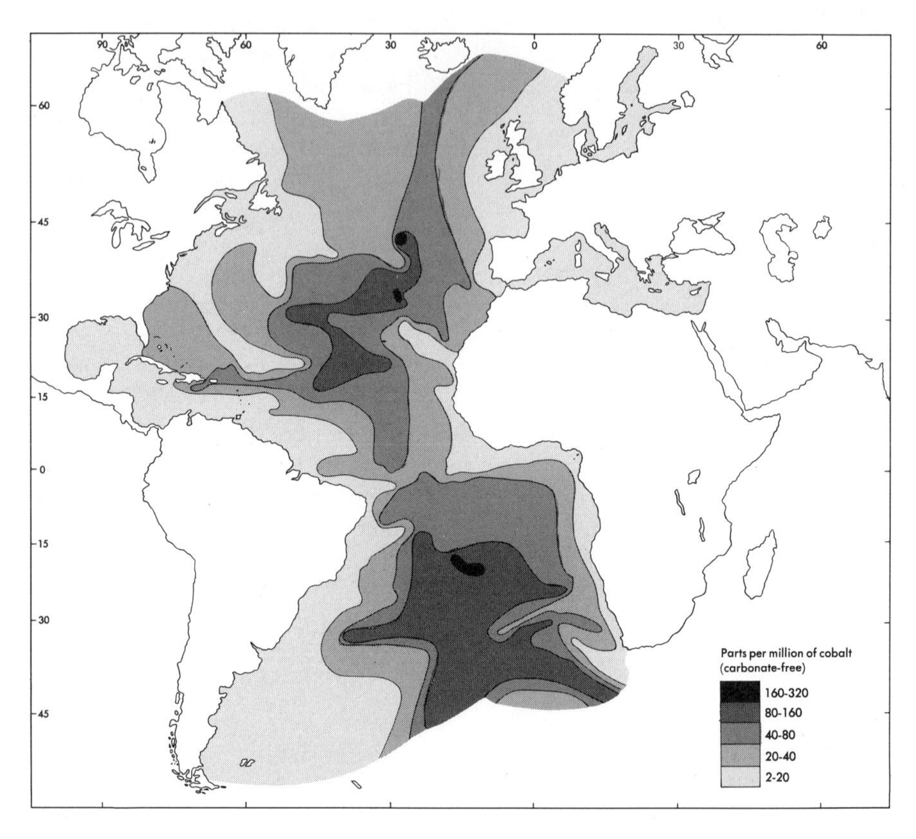

FIG. 6-17 The distribution of cobalt in deep-sea sediments of the Atlantic Ocean (on a calcium-carbonate-free basis) showing that the high concentrations are associated with the Mid-Atlantic Ridge area and the low concentrations are associated with the abyssal plain areas. (After Turekian and Imbrie, 1966.)

The ocean ridges are different environments than the adjacent regions in several ways, all of which may be responsible for accentuating the concentration of metals there: (1) They are the sites of strong submarine volcanism. The release of metals during the interaction of the volcanic rock with surrounding sea water may supply some metals, which are then sequestered into the sediments. It is also possible that hot waters from depth associated with the volcanism can carry metals that end up in the sediments on the ridge. (2) Since the ridges are shallower than the average ocean deep, they accumulate more of the organic material falling from the surface than do adjacent regions. The concentration, for example, of barium and organic matter are higher, on a calcium carbonate-free basis, along the ridges than adjacent regions. (3) The ridge areas, standing high above the ocean floor, do not receive large amounts of bottom-transported detrital material. The fine fraction derived from the continents primarily by atmospheric transport is known to be high in many of the metals found in excess concentration in ridge sediments. The absence of dilution by detrital bottom-transported material maintains the high concentrations received from the atmospheric fallout.

The Red Sea Deposits

The Red Sea is an unusual environment resembling, geologically, the ocean ridge areas except that it is a narrow sea bounded by continental rocks. Deep hot brines have been found in some basins within the Red Sea and underlying them are deposits rich in many metals such as manganese, lead, copper and zinc. The sources of these metals have been shown to be sediments on the adjacent land. Hot salty water flowing through these sediments and erupting into the deep basins in the Red Sea transports these metals and deposits them within the basin.

Such a mechanism is not applicable to the major ocean ridges because they are too far away from continents and the adjacent sediments are too thin and generally lower in topography to permit transport in this fashion.

FIG. 6-18 Distribution of organic carbon and copper in sediments off Walvis Bay, an area of upwelling and high biological productivity. The dashed line is the 500-meter isobath. (From Calvert and Price, 1970.)

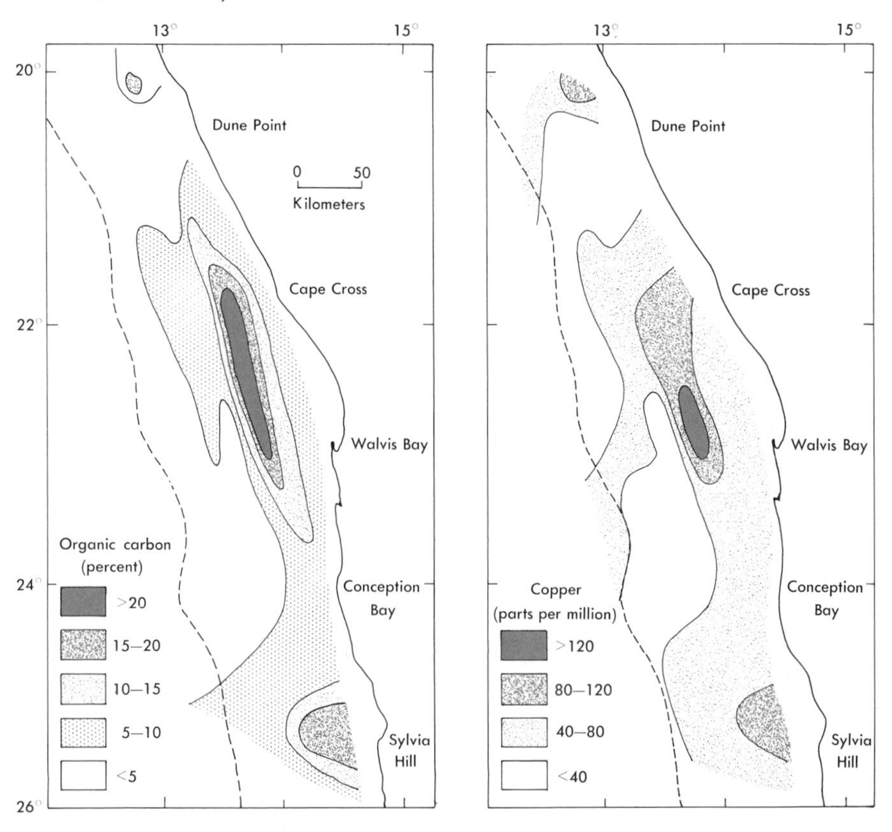

Metalliferous Deposits in Regions of High Biological Productivity

Trace elements may occur in marine organisms as specific compounds, such as copper in hemocyanin in the blood of crustacea, or they may be adsorbed on the phosphatic tests of the zooplankton or embedded as microcrystals in the tissues of microorganisms. In areas of high productivity we should expect transport of trace metals to the ocean bottom. Figure 6-18 shows that the concentrations of the trace metals are indeed high in the sediments of an upwelling area, such as the Walvis Ridge region off southwest Africa. Phosphate concentration is also high in these sediments.

seven

the record in sediments

Sediments accumulate under the force of gravity and form deposits that are roughly parallel with the ocean surface. At any particular location, if the sediments have not been disturbed by slumping, the downward sequence is a continuous sequence back in time. If there are changes in the color, texture, mineralogy, paleontology, or any other property, the sediments are said to be layered. The study of these layered sequences in sediments and the information they contain is called *stratigraphy*.

The object of stratigraphic studies is to relate the sequence of sediments at a particular location to: (1) sequences elsewhere in the ocean in order to determine horizontal variations in sediment pattern at a particular time, (2) worldwide climatic events, (3) oceanic circulation patterns and their variations with time, and (4) events on the continents ranging from mountain building to human activities.

With the advent of methods of absolute dating by means of radioactivity, much of the stratigraphic record can be compared with contemporaneous worldwide events. Radioactive dating also permits us to determine the accumulation rates of the various sediment components.

CORING TECHNIQUES

To study the record in sediments an undisturbed sample must be obtained by coring. The simplest corer is a tube with a weight attached dropped into the sediment. This is called the *gravity* corer. Its major limitation is that it commonly

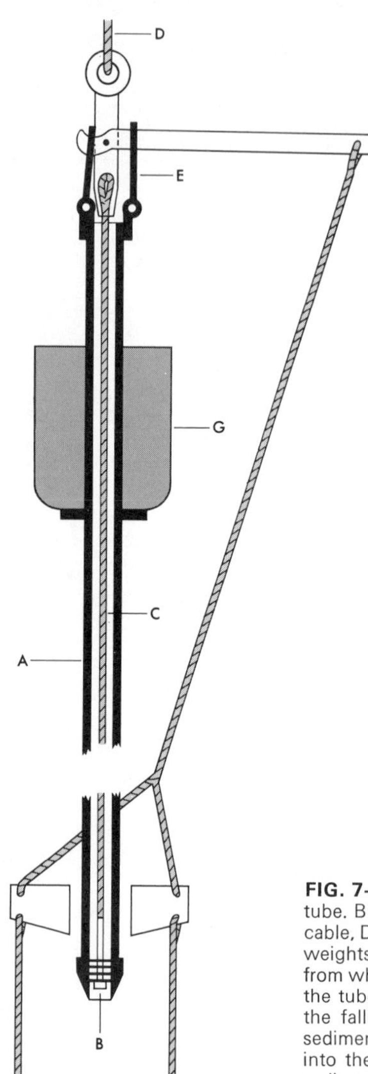

FIG. 7-1 A diagram of a piston corer. A is the coring tube. B is a snugly-fitting piston that is connected to the cable, D, by means of a strong wire, C. When the counter-weights, F, touch the bottom a release mechanism, E, from which the coring tube is suspended is activated and the tube drops rapidly in free fall. The weight, G, gives the falling tube enough energy to be driven into the sediment over the piston. The sediments are sucked into the tube by the piston as the tube penetrates the sediment layers, resulting in a relatively undisturbed stratified section of the sediments of the ocean floor. A piston corer can be used to obtain cores as long as 20 meters in virtually all depths of water. (After Dietrich, 1963.)

can penetrate only less than a meter of the sedimentary pile. A deeper penetration of the sediments is obtained by the *piston* corer (Fig. 7-1). Penetrations to 20 meters are common with this technique.

The problem with both of these corers is that sometimes there has been uncertainty about the recovery of the very top few centimeters at the sediment-water interface. This is mainly due to the fact that the core barrels have generally been of relatively small diameter (about 8 centimeters). This problem has been met successfully in many cases by using either a piston corer of a very much larger diameter or a special type of corer called a *box* corer (Fig. 7-2). The box corer provides a virtually undisturbed record of no more than one meter of sediment but its greater guarantee of the recovery of the sediment-water interface and the large size of the sample make it useful for many problems.

In shallow coastal waters (less than 40 meters) coring can be done by divers. The advantage of this technique is that it provides a good knowledge of the terrain from which the core was obtained. This technique is most useful for short (approximately 30 centimeters) cores specifically to study the very active sediment-water interface.

More recently the powered techniques used to drill for oil on land have been used in the oceans for scientific purposes. The *Glomar Challenger* has been equipped to drill holes in the bottom of the sea. This enterprise, called the Deep Sea Drilling Program, is sponsored by the National Science Foundation and managed by the Scripps Institution of Oceanography. Cores have been taken

FIG. 7-2 The operation of a box corer. (A) The shaft of the box corer approaches the sediment-water interface at the end of a cable, weighted down by over 100 pounds of lead. The jaws of the sediment seal or shovel are open. (B) As the shaft penetrates the sediment, the cable is relaxed. As the cable is pulled up, it pulls on the jaws which slice through the surrounding sediment to seal the shaft. After this is accomplished the box corer is returned to the ship. (Design by A. Soutar, Scripps Institution of Oceanography.)

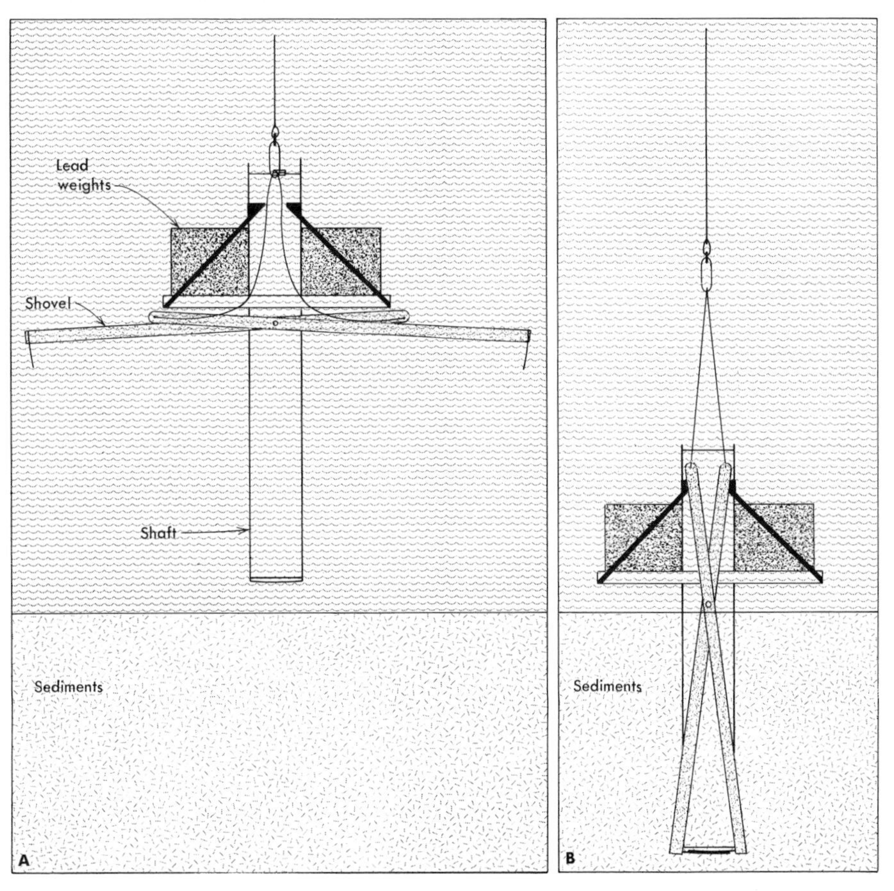

almost routinely in waters as much as 5,000 meters deep, and the recovered cores have shown penetrations as much as 400 meters. The limit of penetration is often set by encounter with a thick volcanic layer through which drilling is difficult. With the new capacity for reentry into a hole, it is possible to continue the drilling to even greater depths by replacing the worn drilling head aboard ship and sending it down again into the same hole.

PRINCIPLES OF RADIOACTIVE GEOCHRONOMETRY

One of the critical pieces of information required to study the record in the cores is to have some measure of the time represented by the strata sampled in the core. Only then can events over widely disparate environments, such as continents and the sea, be correlated and rates be determined for accumulation of sediments or for biological evolution. The discovery of the clocklike nature of radioactivity provided the hope of obtaining this need for telling time in the ocean bottom.

Radioactivity is the result of adjustment of the nuclei of atoms from unstable to more stable states. In the process, energetic particles are ejected from the nuclei; these are called α particles, β^- or β^+ particles, and γ rays. An α particle is a helium nucleus; a β^- particle is a high-speed electron and β^+ is the equivalent antiparticle called a "positron"; and γ rays are electromagnetic radiation, more energetic than X rays. The original, radioactive atom is called the *parent* and the resulting atom is called the *daughter*. A daughter may either be stable or radioactive itself, wherein it becomes the "parent" to the next "daughter" in the decay series. The term *nuclide* is commonly used to describe any observable assemblage, no matter how unstable, of neutrons and protons in a nucleus. *Isotopes* are nuclides having the same atomic number (and thus are the same element) but different numbers of neutrons in the nucleus. The radioactive decay rate of a collection of atoms of a particular unstable nuclide is proportional to the number of atoms present. The proportionality constant is called the decay constant, and the equation for radioactive decay can be written:

$$\text{rate of radioactive decay} = \frac{dN}{dt} = -\lambda N$$

where $\frac{dN}{dt}$ is the incremental change of the number of atoms per incremental change in time
λ is the decay constant
N is the number of radioactive atoms present at that instant

With time the number of radioactive atoms in the collection will be diminishing; since the decay constant itself does not vary for a particular nuclide, it is obvious

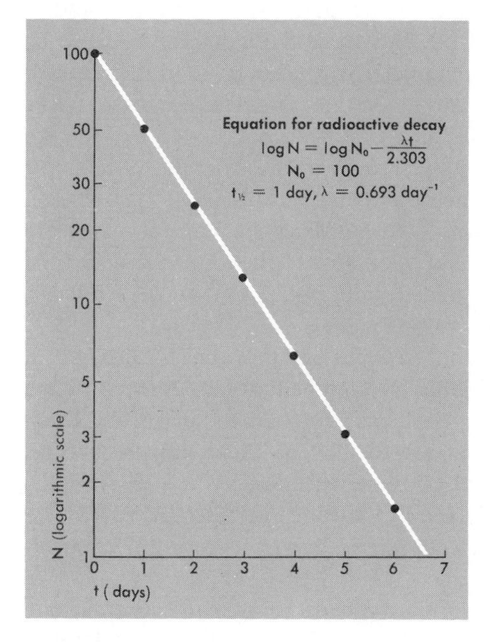

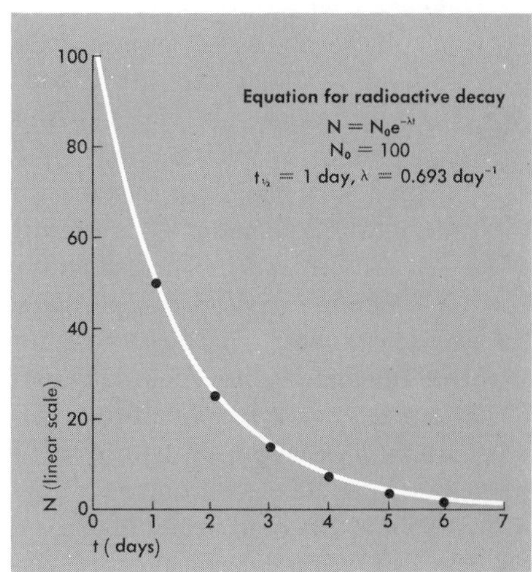

FIG. 7-3 Radioactive nuclides decay to more stable nuclides according to a definite law. This is called exponential decay, and is shown in the right plot. N_0 is the number of original atoms (assumed to be 100 in this example) and N is the number of atoms left after a length of time t has elapsed. In the example the half-life, $t_{1/2}$, is chosen as one day; hence the decay constant, λ, is 0.693 day^{-1}. The left curve is a plot of the logarithm of N against time. Exponential decay when plotted on semilogarithmic coordinates, as was done here, results in a straight line. The conversion factor between natural logarithms and logarithms is given as ln (natural logarithm) $N = 2.303 \log N$. Hence the slope of the line is $-\lambda/2.303 = -\log 2$.

that the rate of decay will also diminish with time. Hence the radioactivity of a collection of atoms at any time is a direct measure of the number of radioactive atoms remaining in the collection at the time sampled.

Figure 7-3 shows that radioactivity decreases *exponentially*, as is predicted by the mathematical solution of the equation we have written above. The *half-life* of a radioactive nuclide is the length of time it takes for the number to diminish by exactly half. It is related to the decay constant by the equation:

$$t_{1/2} = \frac{0.693}{\lambda}$$

where $t_{1/2}$ is the half-life
λ is the decay constant

The value 0.693 is simply the logarithm of 2 to the "natural base" (or base e), or 2.303 times the logarithm of 2 to the base ten.

It is evident that if the radioactive species undergoes decay to a radioactively stable daughter, a measure of the growth of the daughter will also be useful as a time indicator. It is this characteristic property of radioactivity, either the systematic decrease of the parent or the systematic increase of the daughter with time, that makes it useful as a geologic clock.

Three types of natural radioactivity have been used in the radioactive dating of geologic deposits (called "geochronometry"):

1. *Primary:* Nuclides of this type have half-lives so long that they are still present in measurable quantities approximately 5 billion years after the formation of the Solar System. These nuclides are potassium-40, rubidium-87, thorium-232, uranium-235, and uranium-238.

2. *Secondary:* The primary uranium isotopes and thorium-232 decay by a sequence of emission of α particles (doubly charged helium ions emitted by the nucleus), and β^- particles (fast-moving electrons proceeding from the nucleus). Intermediate radioactive daughters that are formed by these transformations have relatively short half-lives. Several of these nuclides, associated with the uranium decay series, whose half-lives are in the range of 100,000 years, have been used in geochronometry of deep-sea cores, namely thorium-230, protactinium-231, and uranium-234.

3. *Cosmic-ray induced:* Nuclides of this type are being made continuously at the present time, principally by the action of cosmic rays. These are relatively short-lived nuclides, which are sustained in the atmosphere and oceans at approximately constant levels by continuous production to make up for the loss by radioactive decay. Examples are carbon-14, hydrogen-3, beryllium-7, beryllium-10, and silicon-32.

Before any of these radioactive nuclides can be used for measuring geologic time, however, certain conditions must be met: (1) The radioactive species used as the clock must be isolated from its daughter at the time of deposition, if growth of the daughter is to be measured or, in a decay series, from its radioactive parent if the decay of the species is to be measured. (2) There must not be migration, in or out of the system, of the radioactive species used as the clock, or of the daughter if its growth is to be used. (3) The half-life of the radioactive nuclide must be suitable for the particular time range under study. It cannot be either too long or too short to be generally applicable.

DATING OF DEEP-SEA CORES

All of the major types of natural radioactivity have been used in dating deposits on the deep ocean floor. We will proceed from methods used to date young deposits to methods that are used to date progressively older deposits. In so doing it will be evident that essentially we will be inverting the sequence of our listing of the natural forms of radioactivity. We will also highlight those methods that have been most generally useful.

Radiocarbon Dating

In 1946 Professor W. F. Libby discovered the natural occurrence of the isotope carbon-14. Since then it has been used for dating by archaeologists, anthropologists, geologists, and oceanographers.

Carbon-14 has a half life of 5,600 years, decaying to nitrogen-14 by the emission of a β^- particle. Before man began producing carbon-14 on a large scale by means of nuclear bombs, the principal agent of terrestrial production was cosmic rays interacting with the atmosphere. As cosmic ray particles (mainly protons traveling with energies measured in *billions* of electron volts) encounter the atmosphere, they interact with molecules to produce, among other products, fast-moving neutrons. These neutrons are slowed down through collisions with molecules in the atmosphere, and by the time they reach about 15 kilometers they are very slow-moving; at this point they are called "thermal neutrons." Thermal neutrons react with nitrogen-14, which is the most abundant nuclide in the atmosphere, to produce carbon-14. The carbon-14 atoms combine with oxygen through a series of steps to form carbon dioxide, in which state it enters the carbon cycle on the Earth's surface.

Scientists have discovered that almost all the neutrons produced by cosmic rays are used up in the atmosphere to produce carbon-14, hence the easily measured production rate of neutrons approximates that of carbon-14; this has been determined to be about 9,800 grams of carbon-14 per year for the whole Earth. Knowing the half-life of carbon-14 and the production rate, we can calculate the level of carbon-14 maintained at the Earth's surface. This value is 8.1×10^7 grams of carbon-14, or 8.29 grams per square centimeter of the Earth's surface. The distribution of carbon-14 in the various reservoirs (see Table 7-1) is roughly the same as the distribution of frequently cycled carbon (that is, carbon other than old fossil carbon in limestones and carbonaceous materials).

It is evident that the main reservoir—and therefore regulator—of carbon-14 is the ocean. Any material in contact with the sea, either directly or through

Table 7-1 Distribution of Carbon-14

Reservoir on the Earth's Surface	Grams of C^{14} per Square Centimeter of Earth's Total Surface
OCEAN (H_2CO_3, HCO_3^-, $CO_3^=$)	7.25
ORGANIC CARBON COMPOUNDS IN OCEANS (MAINLY DISSOLVED)	0.59
BIOSPHERE (LIVING MATERIAL)	0.33
ATMOSPHERE (MAINLY CO_2)	0.12
TOTAL	8.29

After Libby, 1955, Radiocarbon Dating, Wiley

carbon dioxide exchange via the atmosphere, will maintain a constant concentration of carbon-14 relative to the stable isotopes of carbon (carbon-12 and carbon-13).

Once carbon, in isotopic equilibrium with the major reservoir, is removed from exchange with the reservoir by an organism's death or the irreversible formation of plant cellulose or calcium carbonate shells, carbon-14 is no longer added to the system and must decrease with time because of radioactive decay. By measuring the amount of carbon-14 relative to a stable carbon isotope (carbon-12) which does not alter in concentration with time, we can gauge how long the carbon-bearing material has been separated from the reservoir.

We have to assume that the abundance of carbon-14 relative to stable carbon has remained constant in the reservoir by the continuous production of carbon-14 by cosmic rays. The assumption of a constant ratio is reasonable for most dating purposes. But if at any time the cosmic ray flux had changed markedly or if carbon devoid of carbon-14 had been injected significantly into the reservoir (for instance, by the burning of coal and oil, the so-called fossil fuels), then the carbon-14/carbon-12 ratio of the reservoir would have been altered. Such changes, small but measurable, have been noted.

The calcium carbonate tests and organic compounds found in deep-sea sediments can be dated by the carbon-14 method. By normal radioactive counting techniques it is possible to date samples as old as 40,000 years before the present. Radiocarbon dating of deep-sea cores has thus permitted us to measure the duration and end of the last major glacial period. The evidence is that the era of cold climate gave way to the present warmer conditions about 11,000 years ago. On the continents this change was marked by the retreat of the continental glaciers that covered large parts of Europe and North America. The oceans concurrently increased in volume—hence in depth—at a very rapid rate as the ice stored on the continents melted; at the same time the temperature of the surface waters increased.

The differences between glacial and postglacial times were expressed in a variety of ways in the sedimentological record of parts of the deep sea. Some of these have already been discussed. One of the effects was the variation in accumulation rate of sediments in some parts of the oceans as a result of climatically controlled processes. In Fig. 7-4 we see a detailed analysis—by radiocarbon, paleontologic, and chemical methods—of a core raised in the equatorial mid-Atlantic. The core is rich in calcium carbonate, and the nonbiogenic component is mainly pelagic. The radical change in climate 11,000 years ago is tied to a sudden change in pelagic clay accumulation rate from 0.82 grams per square centimeter per thousand years during the glacial time to 0.22 grams per square centimeter per thousand years afterward. The last major glacial event started 26,000 years ago, when the ice had retreated slightly on the continents.

Later we will discuss the use of foraminiferan tests as climatic indicators of the Late Ice Age (or the Pleistocene epoch). We have now seen that the most

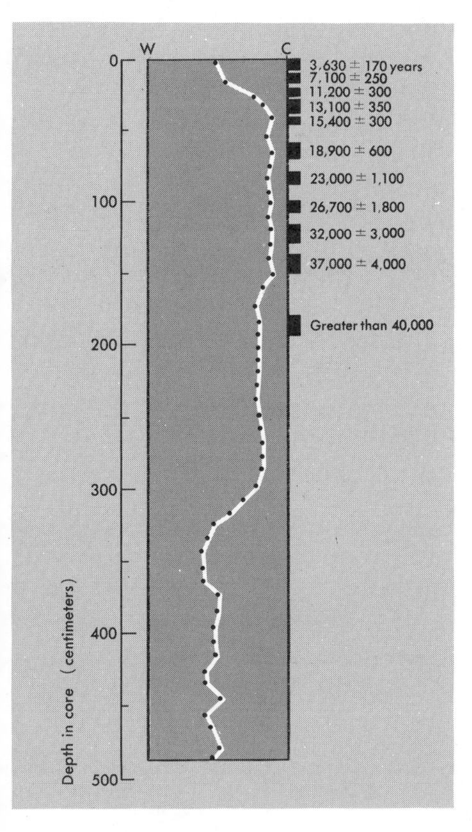

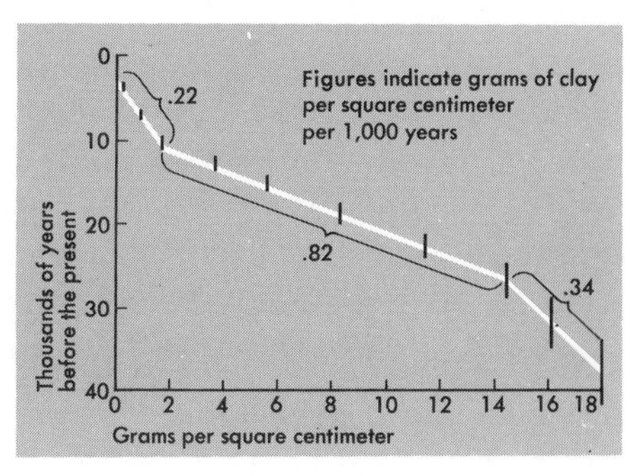

FIG. 7-4 An analysis of an Atlantic equatorial core (A180-74) by radiocarbon dating, chemical, and paleontologic methods. (Left) The climatic curve based on foraminiferan percentages and radiocarbon dates at various levels in the core. (Above) A cumulative plot of calcium-carbonate-free sediment with time. Two breaks are noted. The younger, at 11,000 years before present, marks the end of the last major glacial period; the older, at 26,000 years, corresponds to the time of resurgence of glacial action resulting in the last major glacial period. The rate of clay accumulation is highest during glacial periods and lowest during the nonglacial period. (After Broecker, Turekian, and Heezen, 1958.)

recent change from a glacial climate took place 11,000 years ago and was accompanied not only by biological changes in the oceans, but in places by changes in rates of accumulation. We can thus unambiguously assign a date of 11,000 years before the present for this major event, whenever we find it by the stratigraphic techniques already described. When we know the composition of a deep-sea core down its length and an estimated time for the surface of the core (accounting for mixing by organisms and possible loss during recovery of the core), the additional time point at 11,000 years in the sediment column permits us to determine rates of accumulation of the various components for post-glacial time. We cannot extrapolate these rates to glacial times, however, because of the possibility, as we saw, of radical changes in rate as a function of the different sedimentologic patterns of glacial and nonglacial climates.

Figure 7-5 shows the distribution of postglacial sediment accumulation rates of the nonbiogenic fraction in the Atlantic Ocean. It is evident that the nonbiogenic pelagic component accumulates slowly on the high spots and in areas remote from continents. The abyssal plains have higher accumulation rates generally, principally due to bottom-transported sediments.

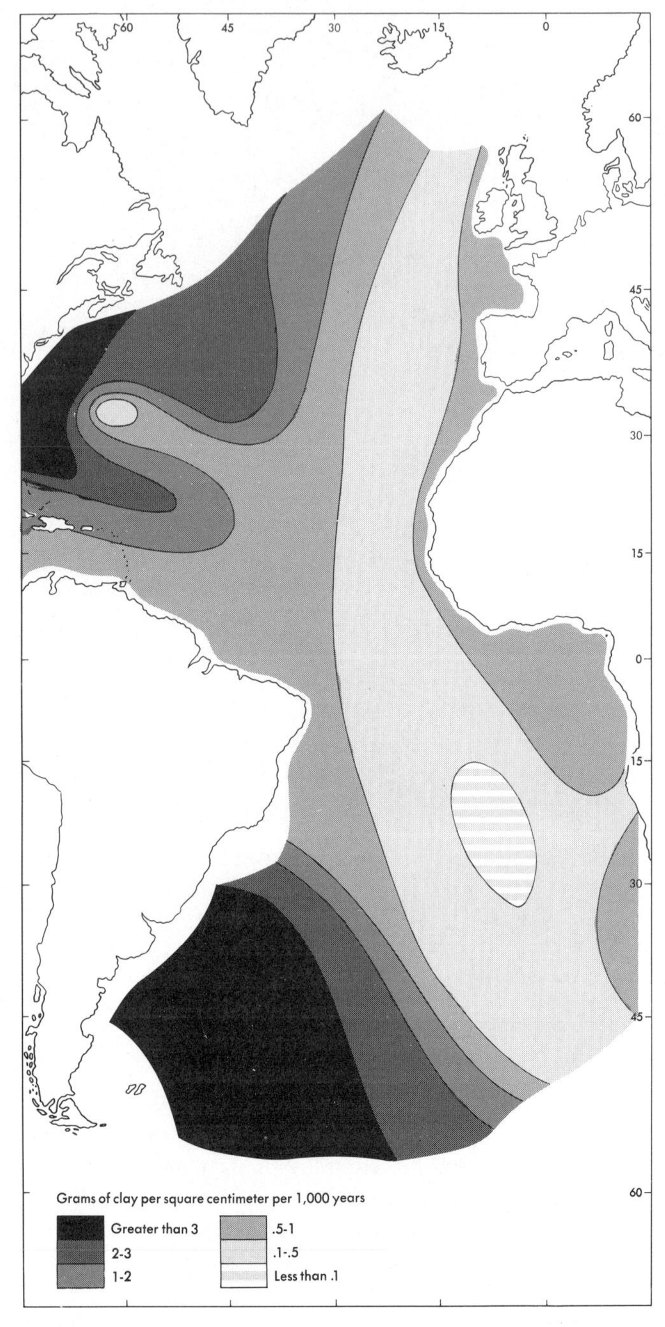

FIG. 7-5 Rates of deep-sea clay accumulation in the Atlantic Ocean during postglacial time. The rates are based on radiocarbon dates and paleontological correlations. The highest rates are in the western Atlantic basins, probably as a result of bottom-transported detritus. The lowest rates are along the topographic highs.

Other Cosmic-Ray-Induced Nuclides Used in Dating

Two cosmic-ray-induced nuclides other than carbon-14—beryllium-10 and silicon-32—have been used in attempting to date deep-sea sediments. Unfortunately, neither is of general applicability and more work remains to be done on the methodology. Berryllium-10 has a half life of 2.5×10^6 years. If we assume a constant rate of deposition in a particular location, then the decay of beryllium-10 can be followed down the length of a sediment core. Knowing the radioactivity of beryllium-10 in the top of the core, we can get from the radioactivity at different depths in the core either a date for a particular point in the core or a rate of accumulation of comparisons are made between different levels in the core. Silicon-32 has a half-life of about 300 years. If at the time of deposition the amount of silicon-32 relative to ordinary silicon incorporated in the tests of siliceous organisms (such as diatoms, sponge spicules, and radiolarians) is assumed to be constant, as with carbon-14, then the decay of silicon-32 can be followed down the core for information on absolute ages or rates of accumulation. The technique is applicable only in areas of rapid silica accumulation, as in the diatomaceous sediments of the Antarctic Ocean.

Uranium Decay Series

Normally, piston coring, with a maximum penetration of 20 meters, can reach through continuous, undisturbed deep-sea sediments to sample the last several million years of record. Radiocarbon dating is limited to the last 40,000 years. For dating the last few hundred thousand years in deep-sea sediments and some other types of marine deposits, the main effort has been centered in the uranium decay series.

The two isotopes of uranium, U^{238} and U^{235}, decay stepwise by the emission of alpha and beta particles (with associated gamma rays) to the stable end products of lead, Pb^{206} and Pb^{207}, respectively. Some of the daughter nuclides near the beginning of the decay scheme can be used in the dating of marine deposits because of their distinctive chemistries and suitable half-lives. The following are the decay schemes for the uranium isotopes showing, underlined, the radioactive daughters that have been used in dating.

$$(1)\ U^{238} \xrightarrow{4.49 \times 10^9 y} Th^{234} \xrightarrow{24.1d} Pa^{234} \xrightarrow{1.18m} U^{234} \xrightarrow{2.48 \times 10^5 y} \underline{Th^{230}} \xrightarrow{7.5 \times 10^4 y} Ra^{226} \xrightarrow{1622y} Rn^{222} \xrightarrow{3.825d} \text{stepwise down to stable } Pb^{206}$$

$$(2)\ U^{235} \xrightarrow{7.13 \times 10^8 y} Th^{231} \xrightarrow{25.6h} \underline{Pa^{231}} \xrightarrow{3.43 \times 10^4 y} \text{stepwise down to stable } Pb^{207}$$

In the dating of deep-sea sediments, the two most important nuclides are Th^{230}, which has a half-life of 75,000 years, and Pa^{231}, which has a half-life of

34,300 years. The basic chemical properties of these two nuclides are very similar to each other and radically different from those of uranium. This can be seen most clearly by comparing the geochemistries of uranium and the very long-lived isotope of thorium, Th^{232}, which has its own decay series. In rocks the ratio of Th^{232} to U^{238} is about four, but this is strongly altered in sea water. The oceans are slightly basic and contain dissolved carbonate and bicarbonate ions, as was implied in the section on carbon-14 dating. In such a solution, thorium is highly insoluble, but uranium forms strong bonds with the carbonate ions, resulting in a highly soluble complex. The concentration of uranium in sea water is thus three micrograms (10^{-6} grams) per liter and that of thorium Th^{232}) is less than 0.0015 micrograms per liter. The ratio of Th^{232} to U^{238} then becomes less than 0.0005 compared to the ratio in rock of *four*.

Since the isotopes of an element all have the same chemical properties, to a very good first approximation, the fate of Th^{230} and the chemically similar Pa^{231} produced in sea water by the radioactive decay of uranium is predicted by the observed behavior of Th^{232}: Th^{230} and Pa^{231} will be removed rapidly from sea water as insoluble phases and deposited on the ocean floor. Hence, these nuclides will be found in deep-sea sediments in excess amounts above that expected in radioactive equilibrium with uranium found in detritus derived from land. There will, however, be no excess of uranium in most deep-sea sediments because of its high solubility due to soluble complex formation with carbonate.

If the excess Th^{230} or excess Pa^{231} accumulates on the deep-sea floor at a constant rate associated with particles that are also accumulating at a constant rate, it is possible to determine the age at any depth in the core by comparing the Th^{230} or Pa^{231} radioactivity at that level with the radioactivity at the top of the core, knowing the half-lives of the nuclides. Alternatively the rate of accumulation can be determined by measuring the rate of decrease of Th^{230} or Pa^{231} radioactivity with depth (Fig. 7-6).

Potassium-Argon Dating

One isotope of potassium, potassium-40, is radioactive with a half-life of 1.24×10^9 years. Of the decay products 88 percent is calcium-40, which is the common isotope of calcium, and 12 percent is argon-40, a rare gas. The argon, trapped in the lattice of potassium-bearing rocks and minerals, accumulates if the temperature of the material is kept below 300°C. By measuring the potassium-argon ratio in a rock or mineral, we can measure geologic time. This method, because of its extreme sensitivity, can be used to date materials as young as about a hundred thousand years and as old as the age of the Earth.

We cannot date the major part of nonbiogenic deep-sea sediments by this method although they are high in potassium concentration, because the detrital clay minerals of the deep-sea sediments contain relict argon derived from weathered and eroded continental material. Recently deposited sediments

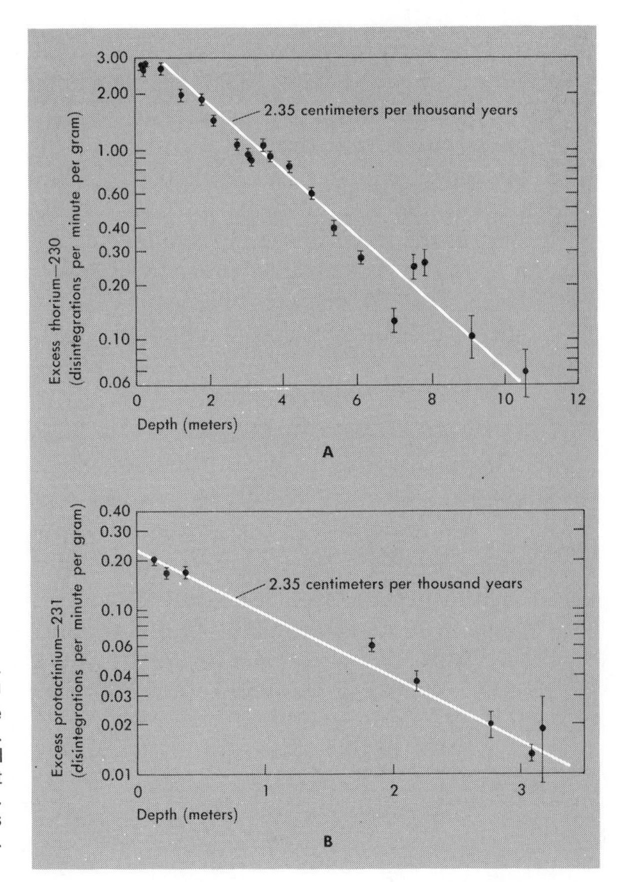

FIG. 7-6 (Right) Plot of logarithm of excess Th^{230} and Pa^{231} with depth in Caribbean core V12-122 (17°00′N, 74°24′W). (A) Th^{230}; (B) Pa^{231}. The vertical line through each point is the extent of analytical errors (one sigma). Note the difference in depth scales between (A) and (B). (After W. S. Broecker and J. van Donk, 1970.)

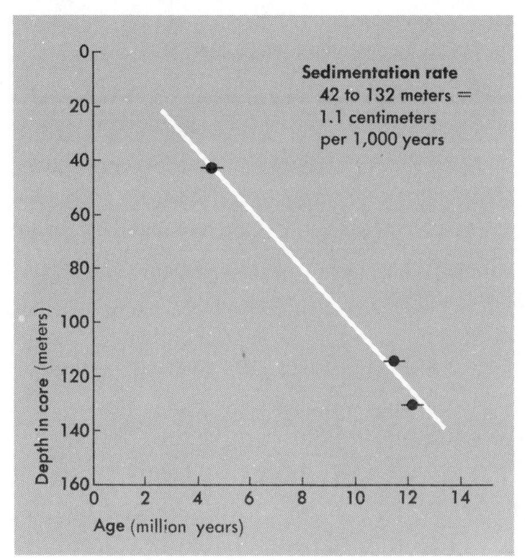

FIG. 7-7 (Left) Potassium-argon dates on glass layers in the core drilled off Guadelupe Island as an experiment in deep-water drilling. In other cores volcanic minerals such as biotite mica and feldspars have been used to obtain dates. (After Dymond, 1966.)

in the deep-sea for instance give "ages" of 200 to 400 million years in the Atlantic Ocean and about 80 million years in the Pacific because the argon was not lost during weathering and erosion.

Associated with deep-sea detrital sediments, however, are volcanic layers. Potassium-argon dates *can* be obtained on these materials because when they were molten all the argon was lost, and any argon measured now had to have formed as the result of the potassium-40 decay. Figure 7-7 shows the analysis of a 150-meter core obtained in the preliminary deep-sea drilling experiment off Guadalupe Island. The rates of accumulation are compatible with rates obtained by the uranium-decay series in the Pacific Ocean.

Magnetic-Reversal Dating

Fairly recently, what appears to be a very useful time-correlative property on a worldwide basis has been discovered—the periodic reversal of the Earth's polarity (Fig. 7-8). It has been observed in volcanic sequences on land dated by the potassium-argon method, that the north and south poles of the Earth's magnetic field have reversed fairly frequently. The evidence is supplied by the strongly magnetic iron oxide minerals in basaltic rocks, which retain the magnetic polarity in force at the time of cooling. For instance, our present polarity began 700,000 years ago, and before that, for about 2 million years, the poles were reversed except for short episodes of "normal" polarity (Fig. 7-9).

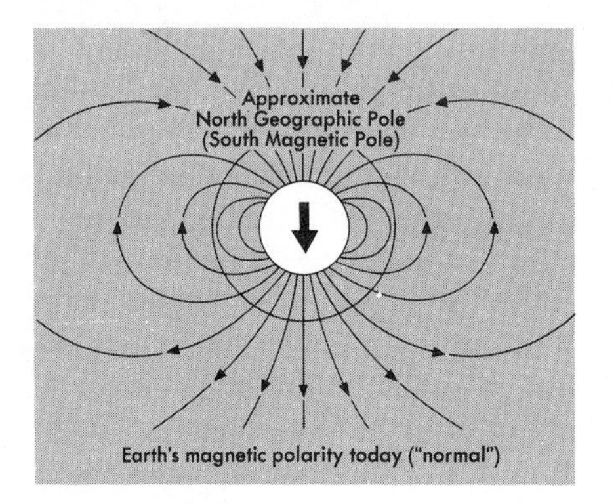

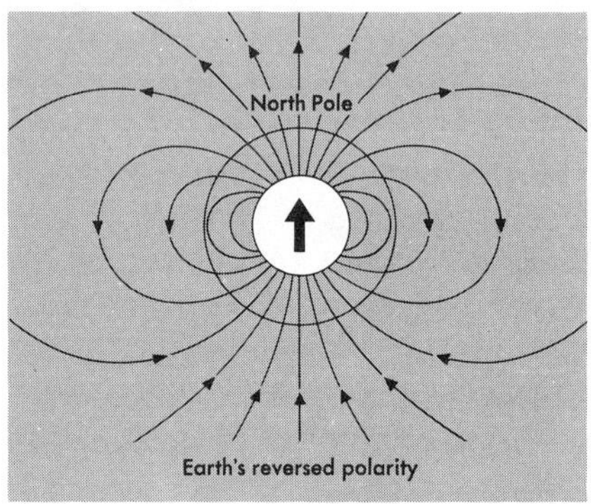

FIG. 7-8 The Earth as a magnet. The north magnetic pole of the Earth at the present time is found in the vicinity of the south geographic pole. That is why the north poles of magnets point north, following the lines of force of the Earth magnet. The reversed polarity (relative to the present) occurs when the north magnetic pole and the north geographic pole are coincident.

The sequence of orientation of the magnetic poles observed in volcanic rocks on land is also recorded by the iron oxide fraction of deep-sea sediments. The orientation of magnets in the Earth's magnetic field can be resolved into a horizontal and a verticle component. That is, the north pole of a freely swinging magnet will point in the horizontal plane to the Earth's south magnetic pole and be inclined vertically in the direction of the lines of force. Because commonly

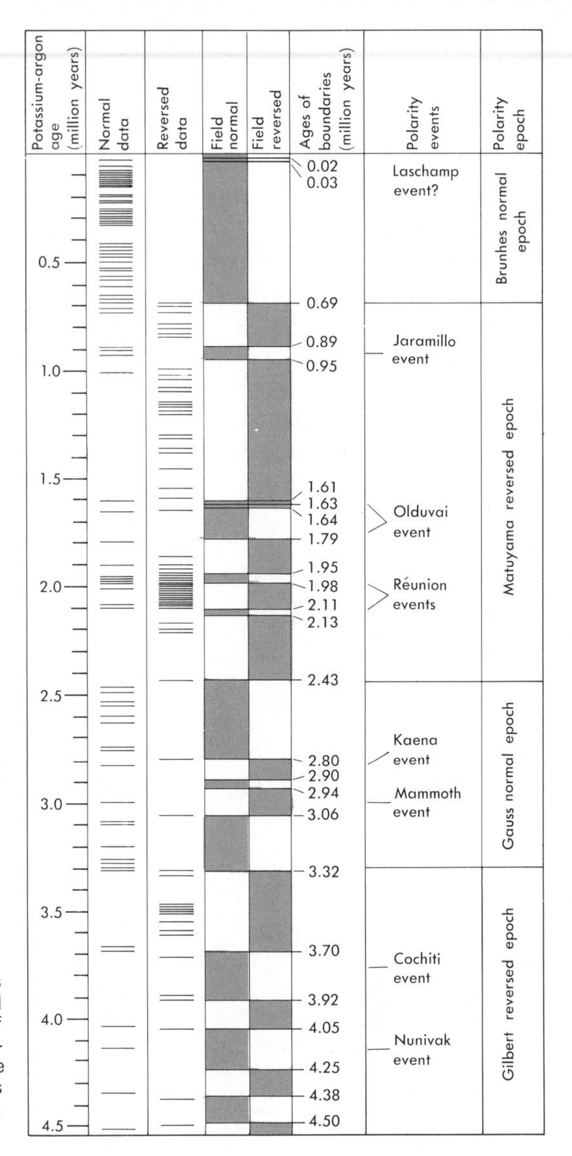

FIG. 7-9 The reversal of the Earth's magnetic field with time was established by determining the magnetic polarity of volcanic materials whose ages were determined by potassium-argon dating. These figures have been established on materials from many parts of the world (After Cox, 1969 with later additions.)

used coring techniques do not give reliable information on horizontal orientation of the core, polarity can be ascertained only where the magnetic lines of force have relatively steep vertical components. The stratification in the core will, however, give a correct indication of the plane of the horizontal. As seen in Fig. 7-8 the vertical component is greatest at high latitudes and approaches zero around the equator. Hence a magnetic vector pointing up, relative to the horizontal stratification of the core, indicates an opposite polarity to a magnetic vector pointing down in the same core. The method cannot be used, however, if there is any question of the continuity of sedimentation in the core or if distinctive paleontologic markers are lacking, since we are dealing with a flip-flop type of record that does not have singular properties.

Summary of Deep-Sea Sediment Dating

Table 7-2 summarizes the techniques discussed above and the ranges and limitations in each method. The fission track technique is still exploratory. To extend the range beyond the 4.5 million-year limit of the radiometrically monitored, magnetic reversal technique, two correlation techniques are employed. One is based on extending the magnetic stratigraphy based on information derived from magnetic anomaly patterns along the ocean floor, as discussed in the next chapter; the other dating method is based on correlation with radiometrically dated fossil sequences obtained in continental areas.

Table 7-2 Summary of Dating Techniques

Method	Feasible Dating Range (years)	Types of Material Datable
C^{14}	0–40,000	carbon-bearing materials—shells, organic matter
Si^{32}	0–2,000	diatom- and radiolarian-rich deposits
Pa^{231}	0–120,000	deep-sea sediments, manganese nodules, corals
Th^{230}	0–40,0000	deep-sea sediments, manganese nodules, corals
K-Ar	60,000 and older	volcanic materials
FISSION TRACKS	older than several hundred thousand years	volcanic materials
Be^{10}	up to 10 million	manganese nodules
MAGNETIC STRATIGRAPHY BASED ON K-Ar	0–4.5 million	deep-sea sediments
MAGNETIC STRATIGRAPHY—EXTENDED CHRONOLOGY	4.5–150 million	deep-sea sediments

CLIMATIC HISTORY FROM DEEP-SEA CORES

In Chapter 3 we saw how the temperature, salinity, and ultimately the circulation of the oceans were coupled to climate. Wind, evaporation, precipitation, freezing, melting—all these influence oceanic properties. The continents obviously are also influenced by the climate and, in an indirect way, transmit the response to the oceans. Water extracted from the oceans can be stored on the continents as ice sheets. This will result in a lowering of sea level. The lowering of sea level by about 100 meters during periods of maximum continental glaciation, as occurred about 18,000 years ago, would expose virtually all of the continental shelves and increase the gradients of streams draining into the sea. The most obvious consequences of this are greater transport of detritus to the deep ocean and the increased saltiness of the oceans.

Marine organisms respond to changes in their environment and this is reflected in the properties of the hard parts accumulating on the ocean floor. With the availability of precise radioactive dating, this allows us to correlate these responses all over the oceans with each other and with continental records as well. Thus a picture of the changing climate of the Earth can be constructed.

Fossil Abundances

The relative abundances of tests of certain species or subspecies of foraminifera, coccolithophorids, radiolaria, and diatoms at different levels in deep-sea cores have been used in relating the marine sediment record to climatic history.

The first successful attempt was in using the percentage of the foraminifera *Globorotalia menardii* in the total foraminiferan assemblage as a measure of warm and cold surface waters. (The higher the percentage the warmer the inferred temperature of the surface ocean.) This method, used in obtaining the climatic curve of Fig. 7-5, was successful in the study of cores dating as far back as 120,000 years from the Atlantic and Caribbean. It does not appear to be a consistently reliable indicator of climate before that time, probably because of the dominance of other temperature-sensitive foraminifera in the total assemblage prior to 120,000 years ago. (As we shall see later, the use of oxygen isotope abundance variations in foraminiferal tests provides a method of inferring general climatic conditions as reflected in the marine record beyond this 120,000-year period.)

Obviously more than temperature has varied in the oceans over time as climate changed. The intensity of upwelling and its resultant nutrient supply rate, the atmospheric cloudiness, the velocity of surface currents, and the salinity also are subject to changes with climate. A general approach to the problem of relating sea water properties, expecially sea surface temperature, and climate is to look at the total assemblage of fossils in the deep-sea sediment record.

If the very tops of undisturbed calcium-carbonate-rich cores are assayed for the relative abundances of the foraminiferal species, a wide range of proportions will be found depending on location. Certain clusters of species increase or decrease depending mainly on latitude which in turn generally reflects sea surface temperature (Fig. 7-10). Once environmentally diagnostic assemblages have been identified, their geographic distribution can be mapped at any particular time in the past using the deep-sea sedimentary record. The exact point in time is established by radioactive geochronometry as discussed earlier. Correlations can then be made along a specific time plane across the oceans.

Since the peak of the last glaciation was about 18,000 years ago, a question of importance to oceanographers and climatologists alike is: What was the surface temperature distribution in the oceans during the peak of glaciation? This question has been answered for the ocean basin most strongly influenced by vicissitudes of glacial advance and retreat—the North Atlantic. Figure 7-11

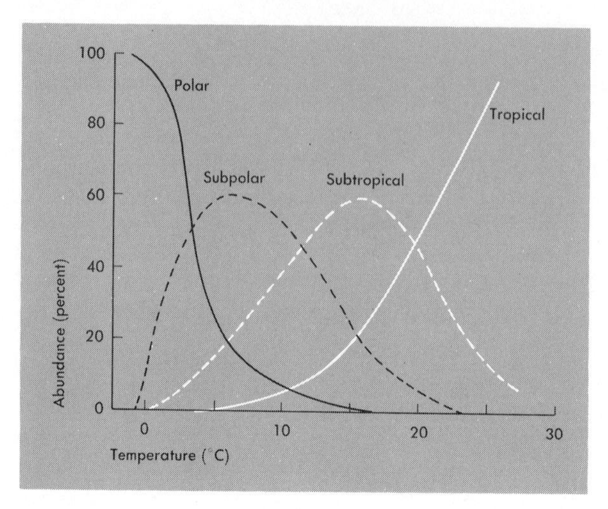

FIG. 7-10 (Left) Idealized abundance trends of foraminiferal assemblages typical of each of four climatic zones. At any temperature of sea water a definite proportion of the four groups is given. (After Imbrie and Kipp, 1971.)

FIG. 7-11 (Below) The observed distribution of foraminifera in core top ("contemporary") sediments (A) in the Atlantic and those at the peak of the last glaciation (18,000 years before present) (B) as expressed by the subpolar cluster of Fig. 7-10. The observed present-day summer ocean surface temperature distribution (C) and the summer ocean surface temperature distribution (D) inferred for the peak of the last glaciation from the foraminiferal distribution patterns. (After N. Kipp and McIntyre, et al. (1976), Geol. Soc. Am. Mem. 145, CLIMAP project.)

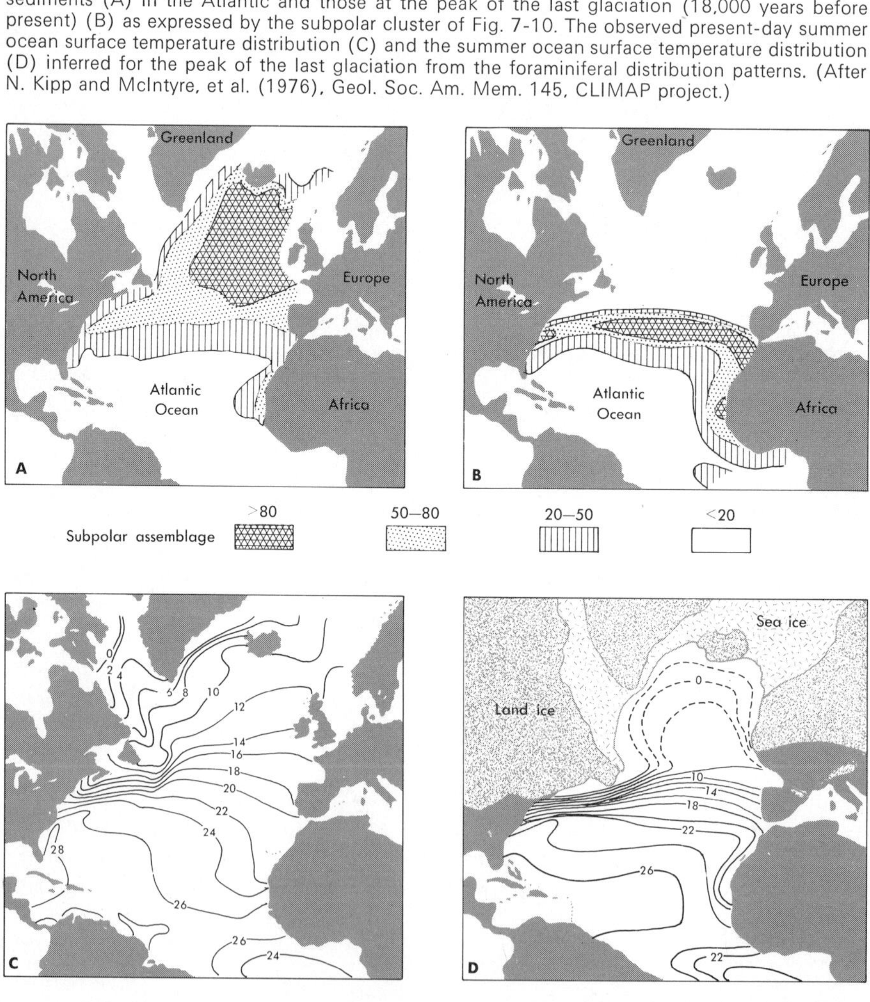

shows how the distribution of diagnostic foraminiferal species at the present time and 18,000 years ago varied and also shows the inferred ocean surface temperature pattern in the North Atlantic at the peak of glaciation.

In principle, with the aid of radioactive geochronometry and long-range correlations, it is possible to construct such maps for the world oceans as far back as satisfactory deep-sea cores are available.

Oxygen Isotopes

One of the most powerful tools for obtaining insights into the ancient climate of the Earth as recorded in deep-sea deposits is the isotopic composition of the oxygen in the tests of foraminifera. This tool has provided information not only on the glacial ages but also the millions of years before that, thanks to the availability of longer time scale sedimentary records through the Deep Sea Drilling Program sponsored by the National Science Foundation.

Isotopes of an element have the same number of protons in the nucleus but different numbers of neutrons. The greater the number of neutrons the greater the mass number of the isotope. Since the differences in mass are small, to a good first approximation the isotopes of an element have identical chemical properties. On the closer view, however, it is apparent that the differences in masses between isotopes of an element to affect chemical reactions, since atoms are being transferred from one molecule to another. Energy differences, as dictated by the masses in motion, result in the unequal distribution of the isotopes among the chemical species (or compounds) involved.

To describe the distribution of isotopes among the compounds when equilibrium is attained, we must use the laws of chemical equilibrium discussed in Chapter 9. Since we are interested mainly in the distribution of the isotopes between two compounds (or phases) as a function of temperature, we can restrict ourselves to reactions involving the simple exchange of isotopes between the compounds. In the case of calcium-carbonate deposition from sea water, an isotope-exchange reaction can be written for the two common isotopes of oxygen, oxygen-16 (99 percent) and oxygen-18 (1 percent), in the following manner:

$$\tfrac{1}{3}\,\mathrm{CaCO_3}^{16} + \mathrm{H_2O}^{18} = \tfrac{1}{3}\,\mathrm{CaCO_3}^{18} + \mathrm{H_2O}^{16}$$

We use fractions of molecules to focus on the exchange of single atoms of oxygen between molecules.

For this reaction the equilibrium constant, K (see Chapter 9), is written:

$$K \text{ (a function of temperature)} = \frac{\text{(the product of the number of molecules of each of the products)}}{\text{(the product of the number of molecules of each of the reactants)}}$$

$$= \frac{(\text{no. molecules } CaCO_3{}^{18})^{1/3} \times (\text{no. molecules } H_2O^{16})}{(\text{no. molecules } CaCO_3{}^{16})^{1/3} \times (\text{no. molecules } H_2O^{18})}$$

At equilibrium, calcium carbonate will have a different ratio of isotopes from that of water because the chemical bonds are different for the two molecules. Although the removal of calcium carbonate from solution as a precipitate will alter the isotopic composition of the water, practically speaking, the oceans are so large and homogeneous that essentially no measurable change in the isotopic composition of sea water takes place by the deposition of the calcium carbonate, so that the ratio H_2O^{16}/H_2O^{18} remains sensibly constant. At different temperatures, different amounts of heavy oxygen will, however, be found in the calcium carbonate relative to sea water. In 1948 Professor Harold C. Urey calculated the expected temperature-controlled variation in the process, the results of which are shown in Table 7-3. This information is the basis of paleotemperature analysis of carbonate shells deposited in ancient seas.

Table 7-3 Oxygen-Isotope Fractionation*

Temperature (°C)	$\frac{O^{18}}{O^{16}}$ Water	$\frac{O^{18}}{O^{16}}$ Crystal
0	$\frac{1}{500}$	$\frac{1.026}{500}$
20	$\frac{1}{500}$	$\frac{1.022}{500}$

* As a function of temperature between water and calcium carbonate; as proposed by Professor Urey in 1948 on the basis of theory and applicable spectral data.

In Table 7-3 the isotopic composition of sea water has been assumed constant; thus the colder the water in which the calcium carbonate forms, the more enriched in oxygen-18 (and thus the "heavier") it is. If the assumption of isotopic constancy of sea water is correct, then a measure of the oxygen isotope composition of a calcium carbonate test deposited from it will be a direct measure of the temperature.

During the process of evaporation of sea water at low latitudes and transport of water vapor to high latitudes, as the general circulation of the atmosphere requires, the sequential precipitation along the way results in a loss of water from the vapor phase richer in oxygen-18 relative to oxygen-16. Thus by the time snow falls on Antarctica or on Greenland it is isotopically about 3 percent "lighter" than average sea water. At the present time, the rate of supply of meltwaters and melting icebergs from these glaciated areas equals the rate of snow accumulation; thus no change in the isotopic composition of sea water occurs.

During the peak of glaciation, however, oxygen-16 is stored preferentially to oxygen-18 in the snow and ice accumulating on the continents. This will increase the oxygen-18/oxygen-16 ratio in the oceans (making it "heavier") at the same time it increases the salinity. The effect would be that any calcium carbonate tests precipitating from the oceans during glacial times would have a higher oxygen-18/oxygen-16 ratio even if sea water had the same temperature as today.

Clearly then, calcium carbonate tests deposited from sea water during glacial times could be enriched in oxygen-18 not only because the surface water was cooler at that time, but because the oceans as a whole were isotopically heavier due to the preferential storage of oxygen-16 in the ice caps. Thus an oxygen isotopic measurement on a calcium carbonate test does not give a simple relationship to temperature but still will be an excellent index of climate.

Can we resolve the contributions of ice storage effect and temperature? Actually, there is a way to do this. If we remember that bottom water in the oceans is produced at the high latitudes in conjunction with freezing processes, we would expect the temperature of this water to show very little response to surface ocean temperatures. Ice formation is the dominant temperature-controlling process around Antarctica. Thus calcium carbonate deposited by bottom dwelling ("benthic") organisms should reflect only the change in the

FIG. 7-12 Comparison of planktonic and benthonic oxygen isotope record from Pacific equatorial Lamont-Doherty core V28-238 (01°01′N, 160°29′E) over the last 120,000 years. The two oxygen isotope variation plots (relative to the same standard) are on the same scale but with the zero scale shifted to the present day planktonic-benthonic difference of 5.3 parts per thousand. The patterns are virtually identical, indicating the dominance of ice storage in varying the oxygen isotope composition of foraminiferal tests in this area of the oceans. (After Shackleton and Opdyke, 1973.)

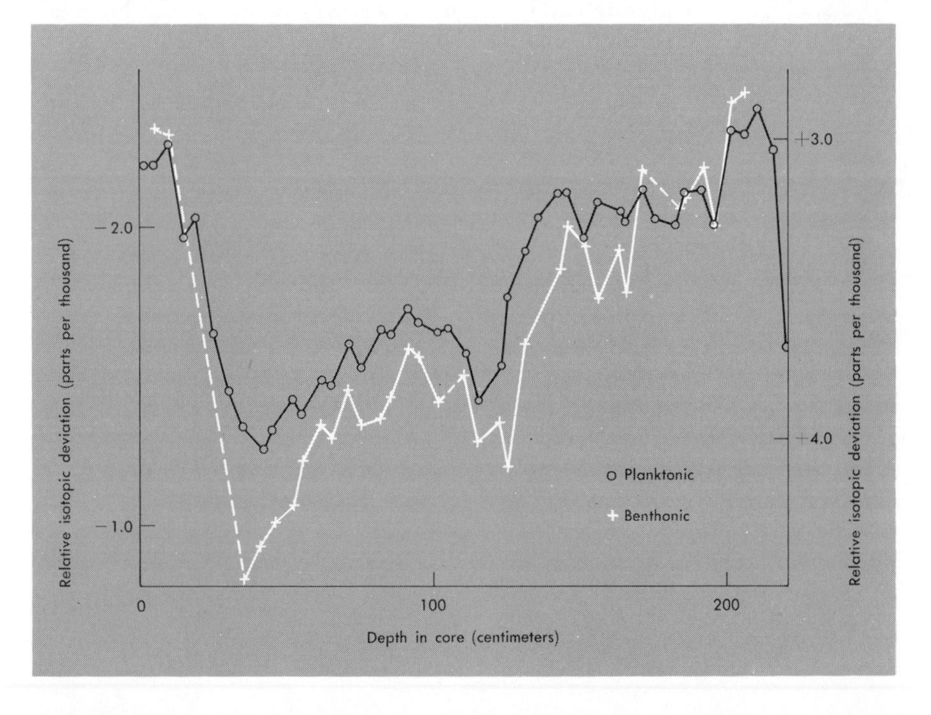

ocean reservoir isotopic composition and have virtually no variation in oxygen-18/oxygen-16 ratio due to temperature variation. Figure 7-12 shows the variation of oxygen isotopes in coexisting pelagic foraminiferans and benthic foraminiferans down the length of a core raised from the equatorial Pacific Ocean. The patterns and amplitude over the last major glacial cycle of 120,000 years are identical, thus indicating that the entire fluctuation in isotopic composition of the pelagic tests is due to changes in the isotopic composition of sea water resulting from more or less continental glaciation. This implies that the temperature of the ocean surface in the equatorial Pacific did not change between present conditions and the peak of glaciation. Indeed, it can be shown by analyzing ice cores from Antarctica and Greenland, which preserve records to about 120,000 years ago, that this result is compatible with the expected preferential storage of oxygen-16 relative to oxygen-18 on the continental ice sheets during glacial ages.

Using the Pacific Ocean data as indicating no temperature variability with time, we can use Fig. 7-12 to extract the water composition effect in other parts of the world ocean. The greatest fluctuation in temperature took place in the Caribbean, with slightly less variation in the equatorial Atlantic. These results are compatible with the paleontologic analysis involved in the construction of the maps of Fig. 7-11.

The oxygen isotope analysis of long cores with a continuous record through the glacial ages can be used to infer the climatic pattern of the past since a heavy isotopic composition implies cold climate accompanied by increased glaciation and a light isotopic composition implies warmer climate and less glaciation. Figure 7-13 shows the pattern for the last 700,000 years. There were about 7 cycles of glaciation during that time, ranging between 80,000 to 120,000 years per cycle. There were also fluctuations before this time but the isotopic documentation is not as complete as for the last 700,000 years.

On the basis of evidence in Antarctica, Australia and New Zealand, continental glaciation of Antarctica started in the latter part of the Miocene

FIG. 7-13 Composite oxygen isotope variation curve from data for planktonic foraminifera from several long cores dating back to 700,000 years (as determined by magnetic stratigraphic dating). There appear to be about 7 cycles of oxygen isotope variations. If these represent 7 cycles of continental ice storage and ice melting, then each full glacial-interglacial cycle takes an average of 100,000 years to complete. (After Emiliani and Shackleton, 1974.)

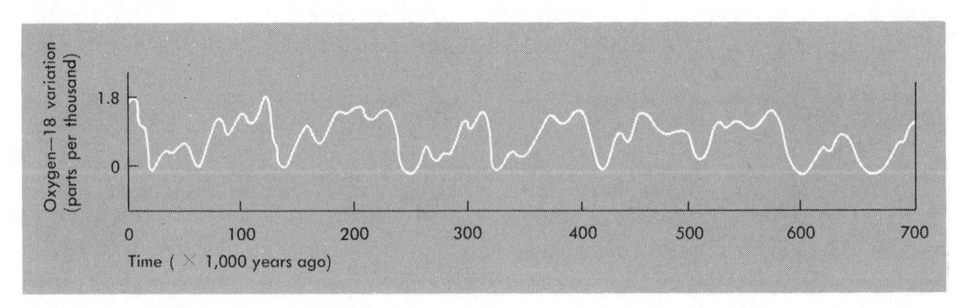

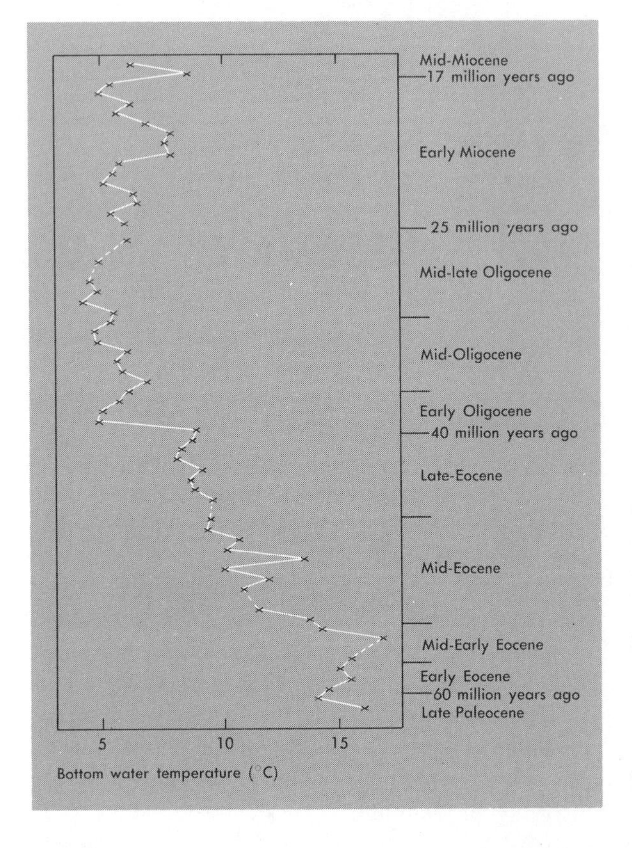

FIG. 7-14 The deep water temperature record between 65 million years ago to about 17 million years ago. During this time there were no major ice caps in Antarctica (or elsewhere). Thus the variation in isotopic composition in benthonic foraminifera was due solely to temperature variations of the bottom water. (From Shackleton and Kennett, 1975.)

epoch and attained the size we observe today, without major alterations, by the end of that epoch about 10 million years ago. Before the beginning of the formation of the Antarctic ice sheet, oxygen isotopic variations in benthic foraminifera reflected not ice storage but actual temperatures of the ocean water. Figure 7-14 shows how the temperature of the ocean bottom at 1,500 meters' depth, south of Australia and New Zealand, varied from the beginning of the Cenozoic era to the establishment of the Antarctic ice sheet. The temperature of the oceans was about 10°C warmer during the early part of the Cenozoic era than at the present. Cooling of the oceans took place fairly rapidly within 20 million years and remained as cool as at present from about 30 million years to today. This implies that a general deterioration of climate occurred prior to the establishment of major ice sheets on Antarctica and later on other continents.

The warm oceans of the early Cenozoic must have resulted in a different chemical regime for the oceans. Everything from the dissolution of calcium carbonate to the aeration of the deep ocean had to be different at that time relative to the present regime.

the history of ocean basins

The unraveling of the history of ocean basins turns out to be the basis for understanding the history of the entire crust of the Earth and indeed possibly of the whole Earth. On land, geologists have been busy studying the interplay of mountain building, erosion, and submergence by the sea as it occurred over and over again during the last 3.7×10^9 years. Large-scale movements of the rocks of the continents by bending and breaking left their record in the terrane accessible to anyone with curiosity, hiking boots, and a geologic pick. Much has thus been learned of the *tectonics*, or large-scale deforming forces that influence the continents.

At one time the ocean basins were thought to be virtually immune from these forces except at the edges where they encountered the continents or within the centers of the basins in the form of volcanic activity. We now know that oceans and continents are all part of a gigantic worldwide tectonism.

OCEAN BASIN EVIDENCE FOR WORLDWIDE TECTONISM

In this section we review all the features of the ocean basin that show its involvement in large-scale movements of the Earth's crust. Many of these features have been known for more than twenty years, but their synthesis into a coherent model of the Earth's crustal evolution is a consequence of the debate and insight occurring since about the mid-1960's.

Continent Fitting

One of the most striking things that one notices on looking at the Atlantic Ocean on a map or globe is that South America and Africa appear as if they are two interconnecting pieces of a jigsaw puzzle separated by water. Greenland also seems to be a piece that fits between Norway and Labrador.

In 1912, long before there was any detailed knowledge of ocean basin structures, this apparent fitting of the continents on either side of the Atlantic Ocean led German meteorologist Alfred Wegener to propose the theory of "continental drift." Wegener had spent time in Greenland and had undoubtedly seen the sea ice and icebergs there drift apart to expose the ocean between them as "leads." He applied this concept to the crust with the less dense continents floating on a more dense substrate and subject to breaking and drifting apart.

Deep-Sea Trenches and Ocean Ridge Systems

In Chapter 2 we noted that certain continental and island margins are marked by trenches rather than a simple continental margin sequence of shelf-slope-rise. These trench areas are always related to adjacent volcanic and tectonically disturbed sequences such as the Andes in South America and island chains like the Aleutians of Alaska or the Japanese Islands (see Fig. 2-2).

Trenches are the deepest holes in the sea and can be maintained only by compressional forces that result in downwarping. It was thought at one time that this was a buckle in the oceanic crust, but the gravity surveys of trenches showed that this could not be the case. As we shall see, the trenches are more likely due to the penetration of an oceanic block either under a continental block or another oceanic block.

The Mid-Atlantic Ridge is part of the world-encircling major ocean ridge system. It differs from the other parts of the system by being mid distance between Europe and Africa to the east and North America and South America to the west. It looks like a suture line along which the continents on either side could be joined.

A central rift valley runs down the axis of the ridge. This type of feature seen on land, in East Africa for example, is attributed to tensional forces that are pulling blocks apart. The median rift valley is not clearly defined in all parts of the world ridge system, but it does exist as a feature in one important segment of it—the Atlantic.

The ridges are not continuously linear. They are offset down their length by breaks. Especially in the Atlantic these breaks are strongly defined by trenches (not to be confused with the often much deeper marginal compressional trenches discussed earlier in this section). These trenches are expressions of a breaking and movement of adjacent blocks in a manner called *transform faulting* (Fig. 8-1). If the blocks were moving with equal probability all along the physical expression of the fault, it would show earthquakes all along this line—such a fault is called a *transcurrent fault*. But if the physical expression is due to both the blocks moving at the same rate away from a line that is offset in the direction of motion, earthquakes will occur only along the fault line between the two offset lines from which horizontal movement is proceeding. This latter case is demonstrated by studying the foci of earthquakes in these regions, and it implies that blocks are moving from a line in both directions.

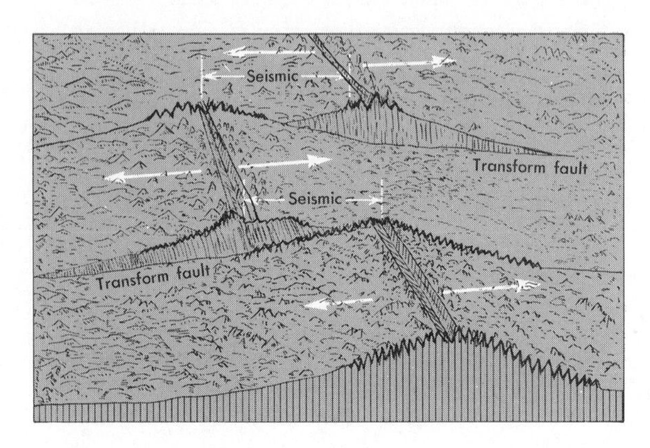

FIG. 8-1 Transform faults. Each block is spreading from the linear apex at the same rate. The dominant earthquakes occur where the relative motion of the blocks is maximum.

The major ocean ridges thus represent features explainable by pulling apart, or *divergence*, and the marginal trenches represent features explainable by pushing together, or *convergence*.

Earthquake Distribution

Earthquakes are not randomly distributed on the Earth. Rather they follow definite linear patterns (Fig. 8-2). The foci of earthquakes are commonly as deep as 800 kilometers, and can be even deeper, but the pattern of deep and shallow focus earthquakes is also strongly patterned.

The results of continuous recording of earthquakes show several patterns: (1) The major oceanic ridge areas are seismically active but all the earthquakes have shallow foci (less than 100 kilometers). (2) The areas associated with deep-sea trenches have the highest seismicity in the world. Earthquakes in these regions show a virtually constant frequency of occurrence from the surface to depths of 800 kilometers. The locus of earthquake foci in these regions describes a plane about 45 degrees from the horizontal descending under the mountain side of the trench (Fig. 8-3). This is one of the reasons for guessing that a trench is made by the movement of one block over the other and enters strongly into the discussion of our model for the tectonic history of ocean basins. (3) Under the major Asiatic mountain ranges along a line westward, including the Himalayan Mountains, frequent earthquakes of both shallow and intermediate depth focus occur.

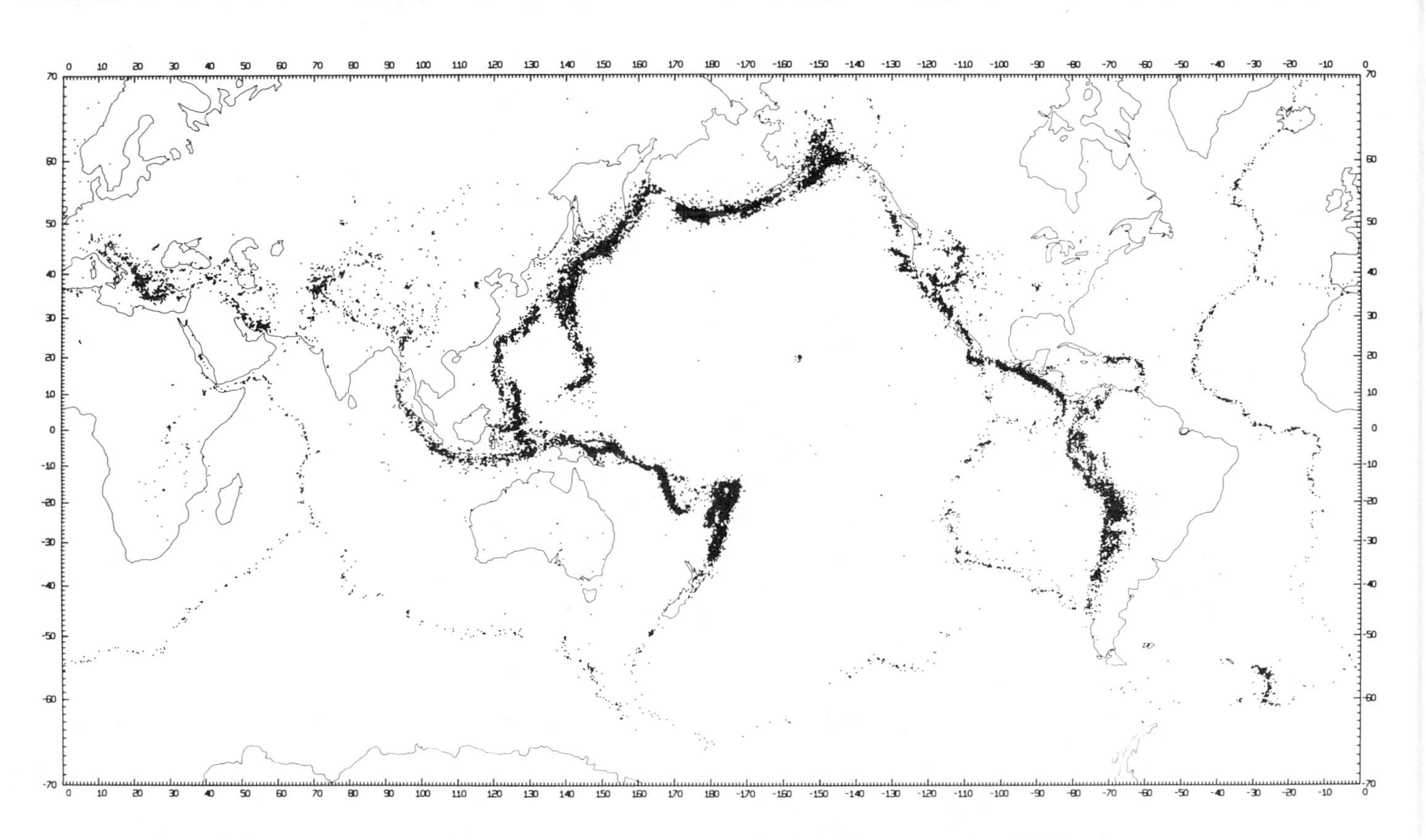

FIG. 8-2 Seismicity of the Earth as recorded between 1961 and 1967 with foci of 0 to 700 kilometers.

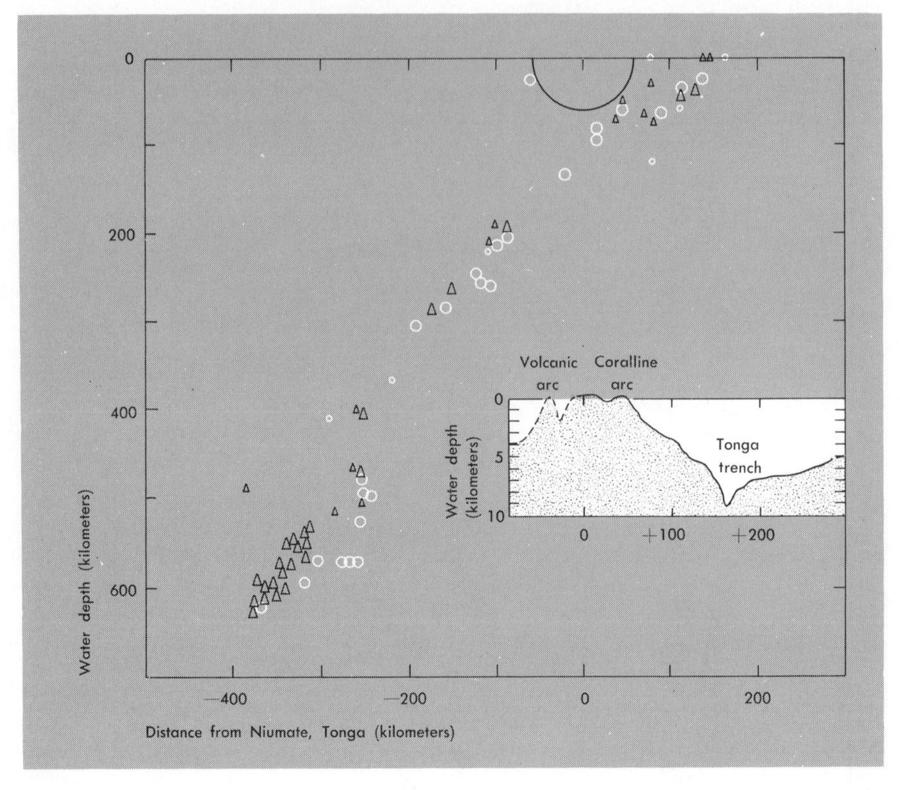

FIG. 8-3 The earthquake foci along trenches. (From B. Isacks, J. Oliver, and L. R. Sykes, 1968.)

Clearly earthquake activity is related to earth motions. We see that the high seismicity associated with various parts of ocean basins implies mobility of the ocean basins in some fashion.

Magnetic-Anomaly Patterns

In Chapter 7 we discussed the fact that the Earth's magnetic field reverses itself periodically. This well-dated phenomenon can be applied to the problem of ocean-bottom mobility.

As has been noted earlier, the major oceanic ridge systems are sites of volcanic activity as well as seismic activity. We know that when a lava flow begins to crystalize on the Earth's surface, the magnetic minerals respond to the magnetic polarity at the time of extrusion. The magnetization is sufficiently strong to survive the vicissitudes of geologic time as long as the temperature is not raised too high again or major chemical alteration does not take place. Lava flows, then, act as a record of the changing magnetic polarity of the Earth. Although clearly oriented volcanic rock samples on land have been used to document the changing polarity of the last 4.5 million years, such an approach is not possible for submarine volcanic material. Dredging volcanic rock samples from the bottom of the sea gives no information on orientation.

A sensitive magnetometer, measuring the Earth's magnetic intensity, towed by a ship or airplane is able to sense the small variations in magnetic intensity over local magnetic bodies such as volcanic rocks. If the magnets have formed in the Earth's present-day polarity, they reinforce the magnetic effect measured by the magnetometer, whereas if the polarity was reversed at the time of the solidification of the lava, the present-day magnetic intensity is subtracted from by the small reversely oriented magnets. Either of these effects result in so called "magnetic anomalies" (Fig. 8-4).

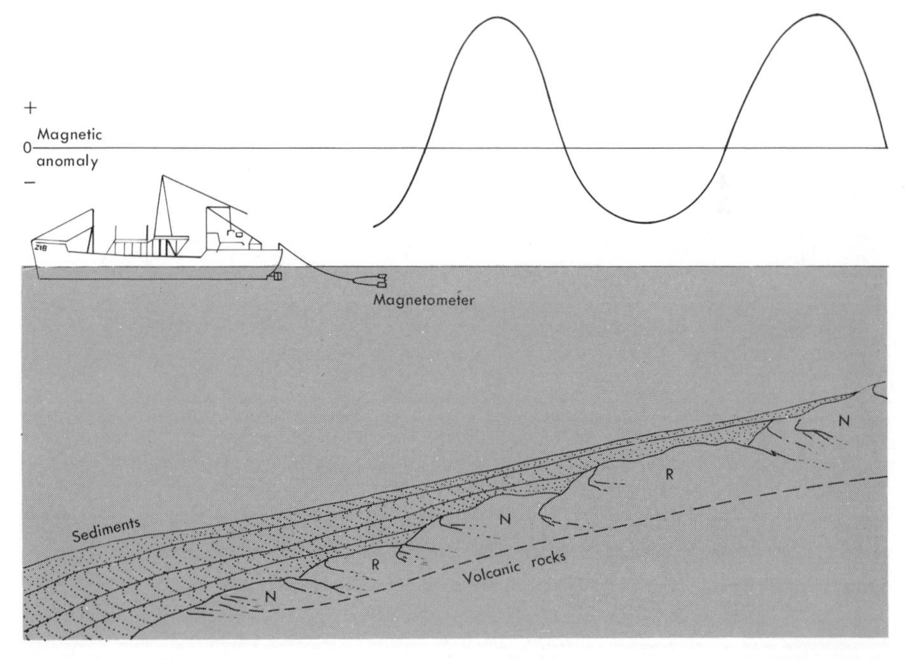

FIG. 8-4 An idealized diagram showing the effect of submarine lava flow magnetic polarity orientation on the magnetic pattern observed with a magnetometer towed by a ship.

Surveys of the major oceanic ridge systems have revealed symmetrical magnetic-anomaly patterns, which are attributed to the processes described above (Fig. 8-5). However, the symmetrical pattern around the center of the ridge implies that if the major volcanic activity has been happening at the crests, then the ridges must be pulling apart to allow the recording of new events by even younger submarine lava flows. From magnetic-anomaly patterns, which can be related directly to known magnetic reversals, it has been calculated that the ridges are separating from either side of the crest at the rate of from 1 centimeter per year (southern Mid-Atlantic Ridge) to 4.5 centimeters per year (southern East Pacific Rise).

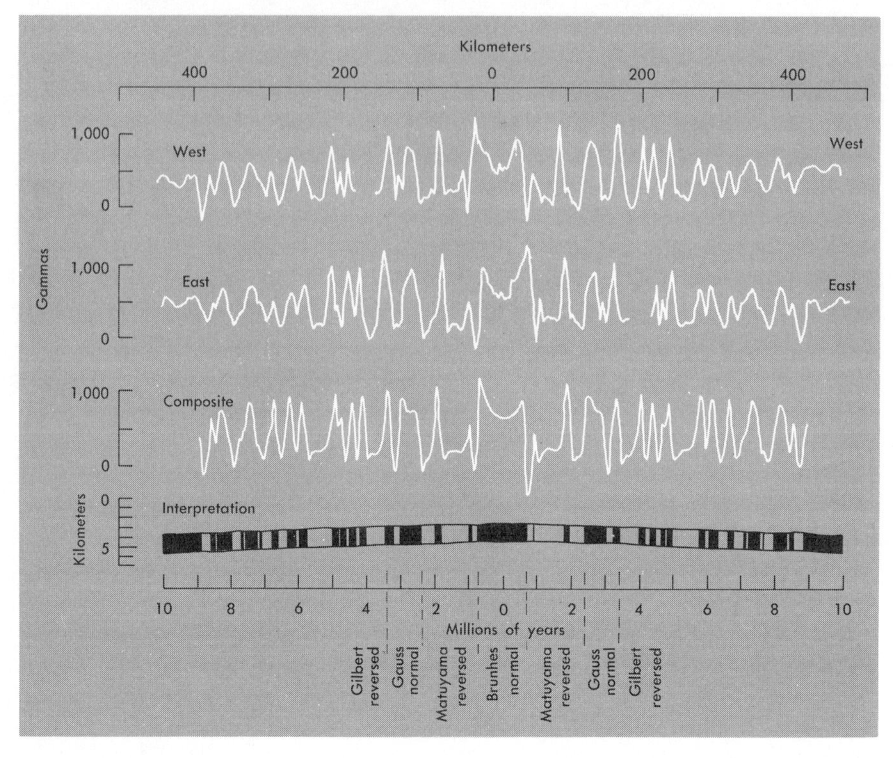

FIG. 8-5 The symmetrical magnetic anomalies around the axis of the mid-ocean ridge systems (here shown for the East Pacific Rise 50°S, 120°W). (From W. C. Pitman and J. R. Heirtzler, 1966.)

This observation is the clearest single evidence for the fact that the ocean floors are highly mobile and indeed spreading away from the ridges.

Sediment Age and Thickness

The oldest sediments sampled from the deep sea by the Deep Sea Drilling Program are no more than 200 million years old, whereas the age of the Earth is 4.6×10^9 years. The probability of finding anything older is very small on the basis of our knowledge of the rates of accumulation of sediments in the deep sea. This fact emphasizes the fact that the ocean bottom is being regenerated through time.

OCEAN FLOOR SPREADING AND PLATE TECTONICS

If we synthesize our knowledge of the properties of the sea floor as described above, we conclude that the oceans are spreading apart along the axis of the major oceanic ridges. Additional divergences occur on the continents as well, but at the present time they are mainly in the ocean basins. Concurrent with the spreading, new volcanic matter is added at the axis of the ridges and this maintains the rugged volcanic topography of the ridge. The emplacement of the volcanic material acts to record the magnetic orientation at the time of emplacement so that a running record of the Earth's polarity is preserved as one progresses away from the ridge axis. The farther away from the axis, the older the basalt. The age of this volcanic "bedrock" of the oceans at any point is thus proportional to the distance from the spreading line from which it was derived and the spreading rate of that divergence area.

It is possible to construct a map of the oceans (Fig. 8-6) by assuming that the rates of spreading measured for the radioactive dated magnetic record of the last 4.5 million years continue into the past at least in some parts of the ocean.

It is obvious that the youngest sediments should be found closest to the ridge axis and the oldest the farthest from the axis. This has indeed been confirmed from cores raised by the Deep Sea Drilling Program.

Ocean floor spreading implies either that the Earth has been expanding to accommodate the new crust generated at the ridges or, for a constant radius Earth, that crust must be obliterated somewhere. On the basis of what we know about the physics of the Earth's interior and the absence of sediments older than about 200 million years anywhere in the oceans, expansion is ruled out. We are then left with finding a way to accommodate the excess old crust as new crust is formed. The answer lies in the convergence areas marked so distinctly in the oceans as the marginal trenches.

As we look at the distribution of convergence and divergence regions visible on the Earth's surface we see that they form the boundaries of large areas of continental dimensions. Some of these areas contain both oceanic and continental areas, some enclose only continental areas, and some only oceanic areas fringed by large islands (Fig. 8-7). These have been called *plates*, and the interaction of the plates at their boundaries is called *plate tectonics.*

The plates are presumed to be about 150 kilometers thick or possibly more, and they include both the crust and outermost part of the mantle. The assemblage of plates, about ten in number, is called the *lithosphere* (see note page 8) by many people active in this area of study. The lithosphere is underlain by the *aesthenosphere* (zone of weakness), which is thought of as the gliding plane for the plate movement. The driving force for the plate motions is energy, mainly from radioactivity, released in the mantle. A sort of convection cell system may be set up that is responsible for the plate motions.

In the original idea of continental drift it was thought that the continental blocks of relatively light or low density material floated on or ploughed through a denser basaltic substratum. It was this concept that for so long delayed the general acceptance of continental drift. Even though the accumulating paleontological and other evidence grew ever more impressive, the physical processes involved seemed to be impossible.

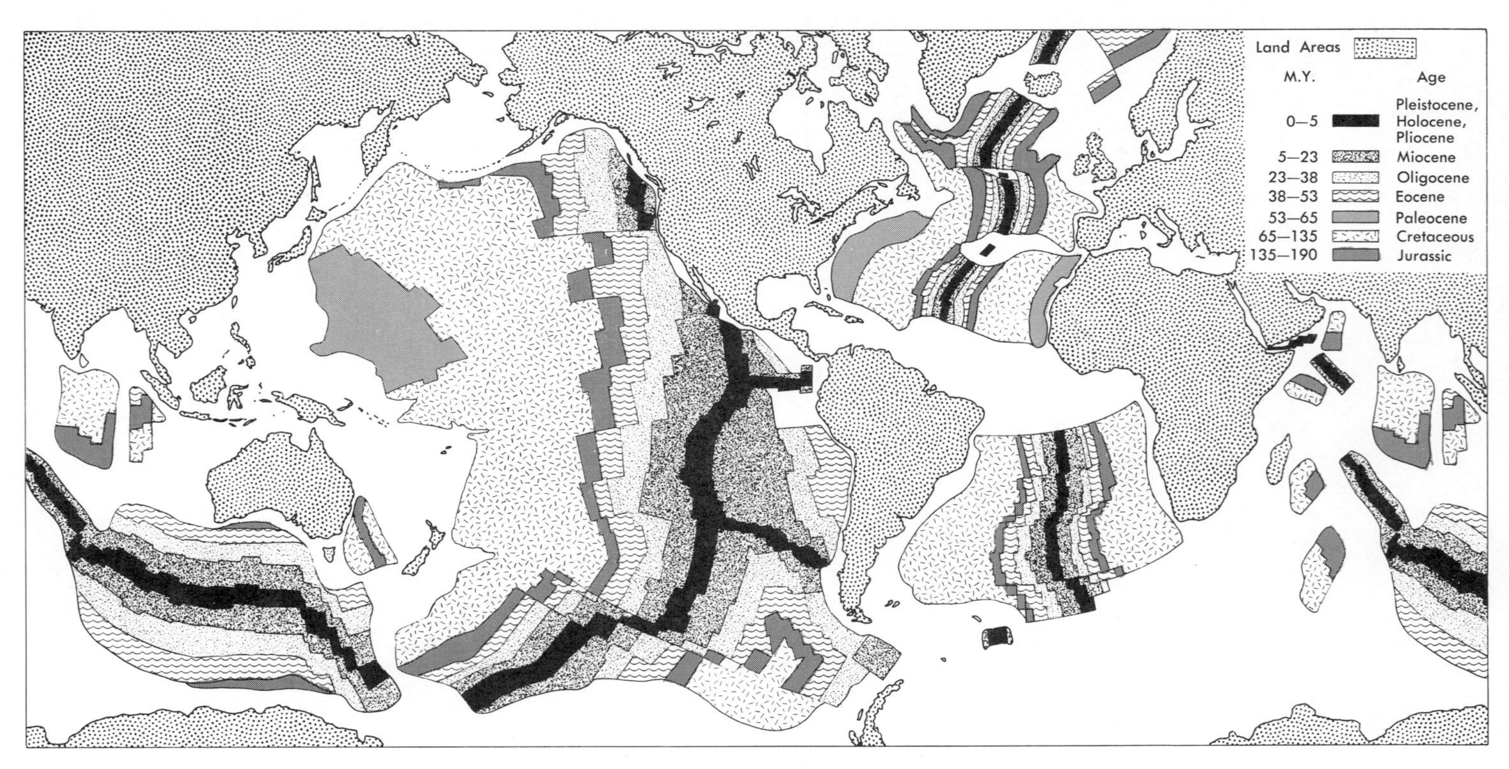

FIG. 8-6 A geologic map of the oceans (age distribution of oceanic crust) based on magnetic anomaly patterns. (From Pitman, Larson, and Herron, 1974.)

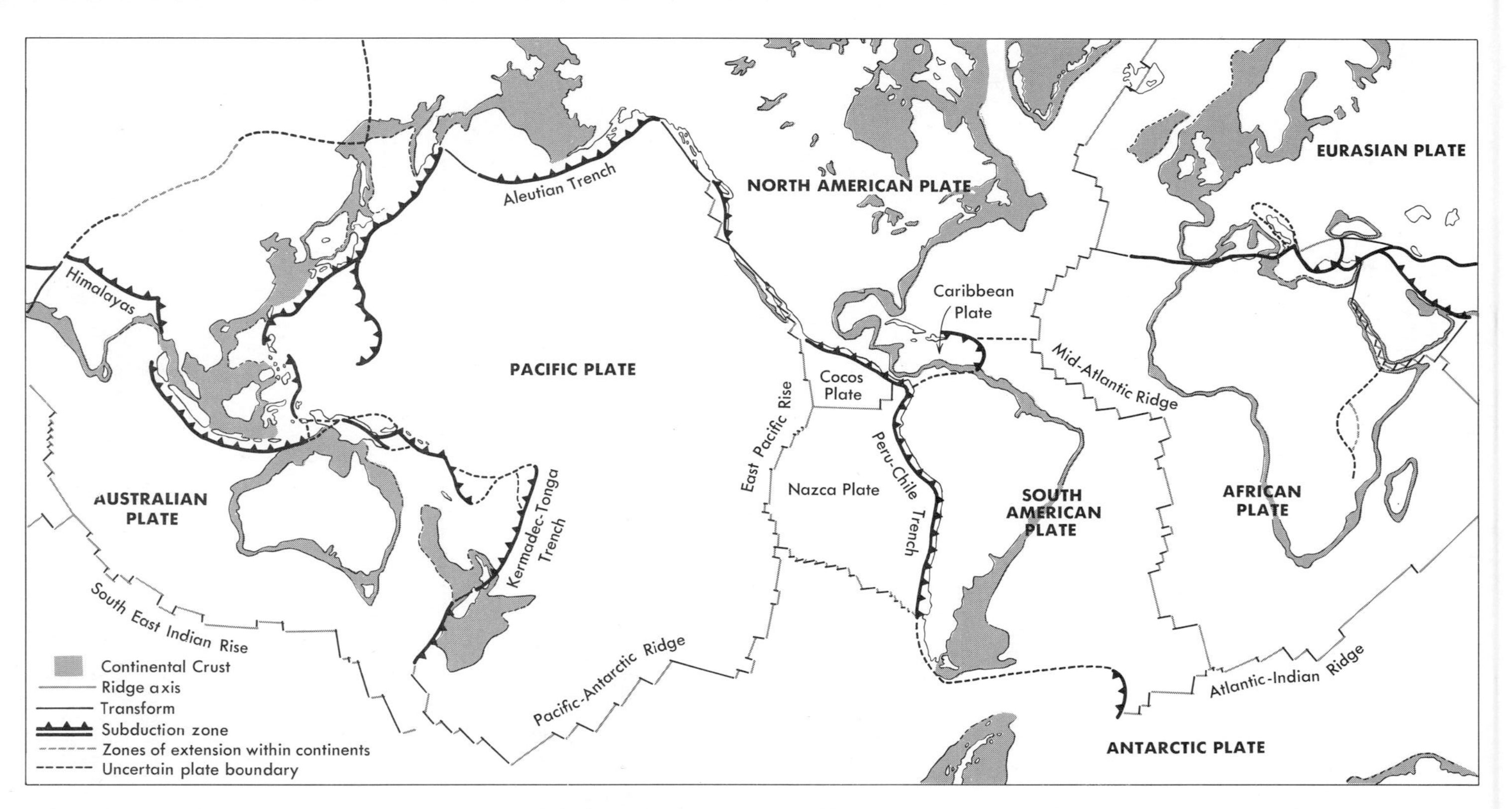

FIG. 8-7 The division of the Earth into "plates." (After J. Dewey, 1972.)

Plate tectonics differs from these early ideas in that the actual moving parts are plates much thicker than the crust alone and carry with them both continental crustal blocks and portions of the oceanic crust (Fig. 8-7).

THE LOCATION OF CONTINENTS AND OCEANS THROUGH TIME

We can determine from the rate of spreading of the oceans on either side of the ridges where the continental features on the plates were in the past. Indeed, if we move backward in time we find that all the continental masses were coalesced about 200 million years ago into one large continent, which has been called Pangaea. We can follow the breakup of Pangaea by reversing the process, and Fig. 8-8 shows this progression.

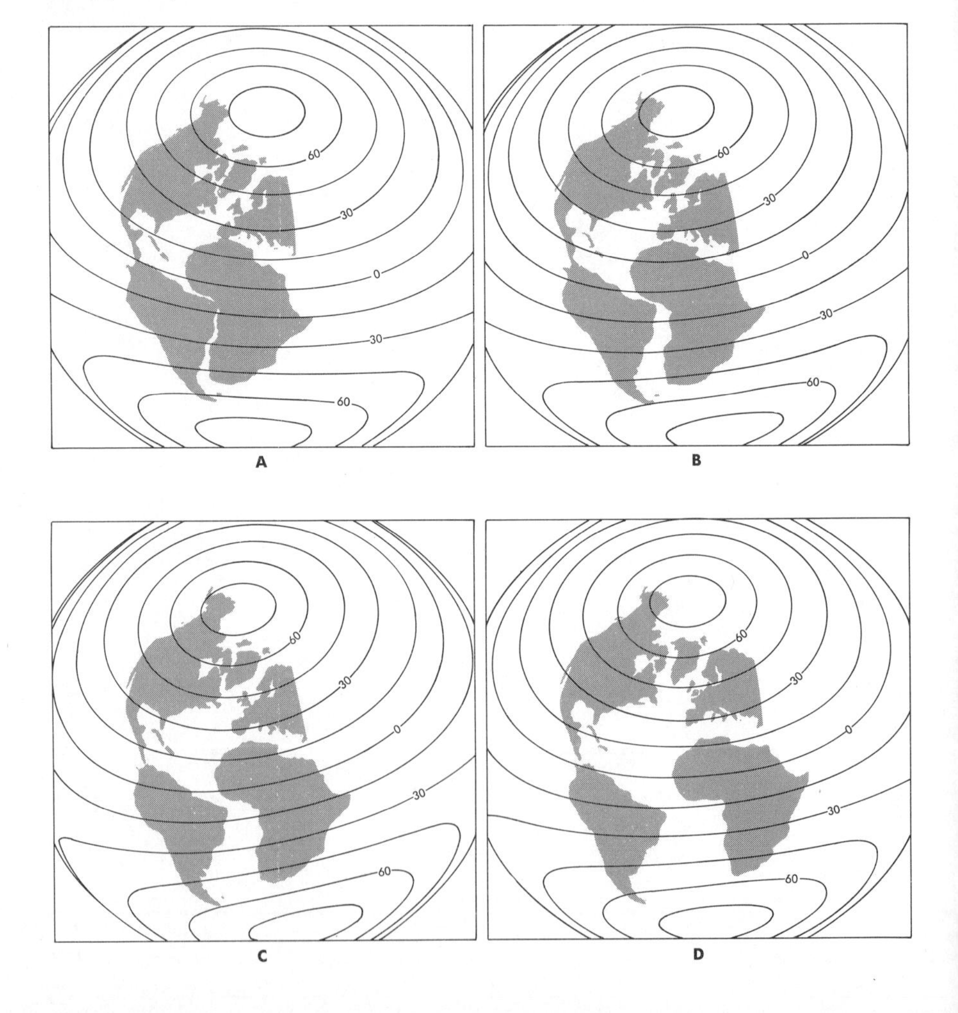

FIG. 8-8 The breakup of Pangaea and the opening of the Atlantic Ocean basin. (A) 150 million years ago; (B) 130 million years ago; (C) 105 million years ago; (D) 65 million years ago. (From Phillips, 1974.)

nine

the chemistry and biology of the oceans

If we have read the paleontologic record correctly, the oceans have been the great incubator for the biological spectrum we see today. The highly diversified ecological niches of the sea and its boundaries have provided the areas of experimentation for many modes of life. From the smallest single-celled species to the most complex mammals, there are adjustments that have been made through evolution to provide adaptation to the light and dark, high pressure and low pressure, cold and warm, salty and brackish, and lush and barren environments of the ocean system.

All organisms are influenced by their environments and in return influence them. They modify the distribution of elements, exude organic compounds capable of influencing other species, and by massive productivity in one part of the system can make graveyards for species in another part. In this chapter we discuss this interplay of life and the oceans and the resultant chemical state of the oceans.

The story actually begins, as we have seen in Chapter 5, with the biologically dominated breakdown of rocks on land that ultimately provide most of the critical chemicals to the sea. We will first discuss the chemical state of the oceans as if it were a solution without the influence of life. Even here, however, we ultimately depend on the boundary conditions, such as an atmosphere composed primarily of oxygen and nitrogen gas, being sustained by life processes themselves. We then proceed to see how the presence of life in the oceans perturbs this system sometimes in surprising and unexpected ways, and finally we try to see how the whole cycle of the surficial Earth is maintained.

THE CHEMICAL STATE OF SEA WATER

In a description of the chemical state of the oceans three properties are of fundamental importance: (1) the concentrations and speciation of the principal ions, (2) the acidity, and (3) the oxidation-reduction potential. If we know what controls these properties, we can then see how they should influence the concentration of the majority of the elements in sea water and the composition of sediments.

Solubility and Ions

If we dissolve 35 grams of household salt (sodium chloride) in one liter of tap water, the resulting solution to a first approximation will resemble sea water. But what does it mean when we say that sodium chloride dissolves? We know that the sodium chloride solution conducts an electrical current very efficiently, whereas pure distilled water does not. Hence, the salt solution must contain charged particles capable of carrying the current, since all electricity involves the movement of electrically charged particles. From this we infer that the sodium chloride has dissociated into charged particles in the solution. We can write the reaction:

$$\mathrm{NaCl\ (solid) + water = Na^+\ (aqueous) + Cl^-\ (aqueous)}$$

The electric conductors in solution are called *ions;* the positively charged ion (Na^+) is called a *cation* since it is attracted to a negatively charged cathode, and the negatively charged ion (Cl^-) is called an *anion* since it is attracted to a positively charged anode.

If we keep stirring more and more salt into our solution, eventually, after a certain point, if the temperature has been kept constant, all subsequently added salt accumulates at the bottom of the container. At this point we say that the solution is *saturated* with respect to sodium chloride. Saturation means that a definite maximum amount of salt will dissolve in a given volume of solvent at a given temperature. This implies an equilibrium state, in chemical terms, and a constant can be defined then to describe this condition. It is called an *equilibrium constant.* In the case of the solution of a salt the equilibrium constant is called the *solubility product constant* (K_{SP}). For a saturated sodium chloride solution, for instance, the K_{SP} is equal to the product of the concentrations of the sodium and chloride ions. (The brackets indicate concentrations.)

$$K_{SP} = [\mathrm{Na^+}] \times [\mathrm{Cl^-}] \text{ (the notation “aqueous” is dropped)}$$

The general rule regarding solubility can now be restated for compounds like sodium chloride as follows: A solution is *saturated* with regard to a particular solid phase if the product of the concentrations of the ions that make up the solid phase is equal to the solubility product constant. The important point here is that the *product* of the ionic concentrations is the critical thing, not the individual ionic concentrations. Concentrations are commonly given in *moles per liter*, a mole being the mass, in grams, equal to the atomic weight of the element. A mole of any two elements has the same number of atoms.

For instance if a salt MX (the term "salt" is used for all solid compounds of ions and is not restricted to common sodium chloride) has a solubility product constant (K_{SP}) of 0.01, then the products of $[M^+] \times [X^-]$ shown in Table 9-1 will all be saturated solutions with respect to the compound MX.

Table 9-1 Saturated Solutions of Salt MX ($K_{SP} = 0.01$)

$[M^+]$	$[X^-]$	$K_{SP} = [M^+][X^-]$
10	0.001	0.01
1	0.01	0.01
0.1	0.1	0.01
0.01	1	0.01
0.001	10	0.01

With a logical extension of the solubility product constant calculation, we can apply it to salts of diversely charged cations and anions.

Normal sea water is undersaturated with regard to all of the major salts that can be formed from the combination of the major ions, except for calcium carbonate in surface waters. That is, the solubility product constants of such salts as calcium sulfate ($CaSO_4$), magnesium sulfate ($MgSO_4$), potassium chloride (KCl), sodium chloride (NaCl), to name some of the more important possibilities, are not exceeded in normal sea water. If we increase the concentrations of the ions by evaporation, various salts will precipitate as the solubility product constant of each is exceeded.

It is evident that the concentrations of the major components of sea water, in particular the anions, set the maximum concentration levels of the minor and trace elements on the basis of solubility product constants of their most insoluble compounds.

Many of the elements found in trace quantities in sea water can form complexes with the dominant anions, thereby increasing their solubility above what is predicted by the solubility product constant of the most insoluble salt. Hence, although silver chloride is extremely insoluble in distilled water, its solubility increases in a sodium chloride solution by several orders of magnitude. Even though the concentration of Ag^+ is diminished by the increased chloride ion concentration, the concentrations of chloride complexes such as $AgCl_2^-$ are high enough to increase the total amount of dissolved silver.

Salts generally have larger empirical solubility product constants in saline water than in distilled (or stream) water even if special complexes are not formed. For instance, the empirical solubility product constant of barium sulfate is 10^{-10} in distilled water (or streams) but about 10^{-8} in sea water. As it happens, the sulfate-ion concentration of sea water is about 100 times that of streams, hence the barium concentrations for sea water and streams should be about the same to give empirical solubility product constants of 10^{-8} and 10^{-10} respectively. Table 9-2 gives a list of the most insoluble compounds of some trace

Table 9-2 Solubility Controls on Concentration of Trace Elements in Sea Water

Element	Insoluble Salt in Normal Sea Water	Expected Concentration (log moles per liter)	Upper Limit of Observed Concentration (log moles per liter)	Expected Concentration in Sulfide-Rich Sea Water (log moles per liter)
LANTHANUM	$LaPO_4$	−11.1	−10.7	—
THORIUM	$Th_3(PO_4)_4$	−11.8	−12.7	—
COBALT	$CoCO_3$	−6.5	−8.2	−12.1
NICKEL	$Ni(OH)_2$	−3.2	−6.9	−10.7
COPPER	$Cu(OH)_2$	−5.8	−7.3	−26.0
SILVER**	AgCl	−4.2	−8.5	−19.8
ZINC	$ZnCO_3$	−3.7	−6.8	−14.1
CADMIUM	$CdCO_3$	−5.0	−9.0	−16.2
MERCURY**	$Hg(OH)_2$	+1.9	−9.1	−43.7
LEAD**	$PbCO_3$	−5.6	−9.8	−16.6

* The expected concentrations at 25°C are calculated on the basis of the following thermodynamic concentrations, *a* ("activities") of the anions: $\log a_{PO_4^{\equiv}} = -9.3$, $\log a_{CO_3^{=}} = -5.3$, $\log a_{OH^-} = -6$, $\log a_{S^=} = -9$.

** Form strong chloride complexes.

elements in sea water together with the maximum concentrations of the trace elements compatible with the major anionic composition of sea water. For comparison, the upper limit of observed concentrations of these elements is also given. It is evident that, on this basis, the oceans are generally undersaturated with regard to the trace elements as well as the major elements except possible lanthanum (and the other rare-earth metals). We shall return to this problem in a later section.

The Acidity of the Ocean

The acidity of a solution is determined by the concentration of hydrogen ions (H^+). The water molecule, composed of hydrogen and oxygen, can form ions by dissociation much as the inorganic compounds or salts do when put into solution. For salts like sodium chloride the dissociation is complete for all practical purposes. The degree of dissociation of water, however, is far from complete. The dissociation of water is written:

$$H_2O \text{ (liquid)} = H^+ \text{ (aqueous)} + OH^- \text{ (aqueous)}$$

The equilibrium product constant for this reaction is written in a manner similar to that for the solubility product constant:

$$K_{\text{dissociation}} = [H^+][OH^-]$$

In pure water, for every hydrogen (H^+) ion formed by dissociation, a hydroxyl (OH^-) ion must be formed; the concentration of each is the same. The value of $K_{\text{dissociation}}$ at atmospheric pressure and 25°C is about 10^{-14}, hence the *hydrogen ion concentration* of pure water is 10^{-7}.

Commonly, purely as a matter of convenience, the hydrogen-ion concentration scale is converted to logarithmic form (base 10) to avoid writing exponents. Such a scale is called the *pH scale:* pH is defined as the negative logarithm of the hydrogen ion concentration. A pH of 7 corresponds to a hydrogen-ion concentration of 10^{-7} and is the pH of pure water. A lesser pH value is considered acidic, a greater one is basic. Since the pH of sea water is around 8, the ocean can be considered as a slightly basic solution.

In the laboratory, the pH of water can be lowered by adding an acid, such as hydrochloric acid or acetic acid (vinegar), since on dissociation it adds hydrogen ions without adding hydroxyl ions; pH can be raised by adding a base, such as sodium hydroxide (lye) or ammonium hydroxide (ammonia water).

The chemical species involved in the regulation of the pH of ocean water are also common compounds: carbon dioxide in the air, carbonic acid, and calcium carbonate. Carbon dioxide gas is the respiratory waste of animals and the starting point for photosynthesis in land plants. When carbon dioxide is bubbled through water, it dissolves to form carbonic acid, thus lowering the pH of the solution. This carbonic acid will react vigorously with calcium carbonate, with an increase in the pH. The reaction is:

$$H_2CO_3 + CaCO_3 = Ca^{++} + 2HCO_3^-$$

It is evident that hydrogen ions from the dissociation of the carbonic acid are in part used up to form bicarbonate ions (HCO_3^-) from carbonate ions ($CO_3^=$) that are derived from the dissociation of the $CaCO_3$. This action tends to regulate the pH of the solution, hence we say the solution is "buffered." More precisely, the series of chemical reactions described above and their intermediate steps can be expressed in the following equations, shown with their associated equilibrium constants (including the dissociation constants and solubility product constants) at 25°C.

$$K_P = 10^{-1.47} \quad CO_2 \text{ (gas)} + H_2O \text{ (liquid)} = H_2CO_3 \text{ (aqueous)}$$

$$K_1 = 10^{-6.4} \quad H_2CO_3 \text{ (aqueous)} = H^+ \text{ (aqueous)} + HCO_3^- \text{ (aqueous)}$$

$$K_2 = 10^{-10.3} \quad HCO_3^- \text{ (aqueous)} = H^+ \text{ (aqueous)} + CO_3^= \text{ (aqueous)}$$

$$K_{SP} = 10^{-8.3} \quad CaCO_3 \text{ (solid)} + H_2O \text{ (liquid)} = Ca^{++} \text{ (aqueous)} + CO_3^= \text{ (aqueous)}$$

$$K_W = 10^{-14} \quad H_2O \text{ (liquid)} = H^+ \text{ (aqueous)} + OH^- \text{ (aqueous)}$$

On the basis of these equations and the dissociation of water, we can relate the pressure of carbon dioxide in the atmosphere to the acidity of sea water on the basis of equilibrium with calcium carbonate, which is given by the following simplified equation:

$$P_{CO_2} = (K_p K_1)^{-1} \left(\frac{2K_{SP}}{K_2}\right)^{\frac{1}{2}} [H^+]^{\frac{3}{2}}$$

The present-day pressure of carbon dioxide is 0.0003 atmospheres, corresponding to a pH of about 8 for sea water, the value which is, in fact, found. Rather large changes in the carbon dioxide pressure result in only small change in the pH.

An interesting question arises from the relationship between the pressure of carbon dioxide in the air and the pH of sea water: Is the carbon dioxide pressure determining the pH or is the pH determining the carbon dioxide pressure?

The argument that the carbon dioxide pressure is independently determined, and thus influences the pH of sea water, is based on the premise that it is the maintenance, by respiration, of the supply of carbon dioxide in the atmosphere that is the critical factor. Hence, it is argued, as the result of the balance of life processes producing and using carbon dioxide, a steady-state level of carbon dioxide is maintained in the atmosphere.

The alternative view is that reactions of silicates in the oceanic sediments control the hydrogen ion concentration. This control can be effected either by direct reaction of hydrogen ions with the clay minerals so as to transform them from one kind to another or by simple exchange with the cations at adsorption sites in the clay minerals. To understand the basis of these reactions in controlling the acidity of sea water, we must understand the mode of origin of the clay minerals in sediments. These are ultimately the products of weathering as was indicated in Chapter 5. We can write a generalized weathering reaction in the following way:

$$Na^+ \text{ feldspar} + H^+ \text{ (aqueous)} = \text{kaolinite} + Na^+ \text{ (aqueous)} + SiO_2 \text{ (aqueous)}$$

and an equilibrium constant can be written for this reaction:

$$K = \frac{[Na^+][SiO_2]}{[H^+]}$$

In the oceanic realm it is presumed that the same reaction is also occurring, but mainly in the reverse direction—that is, kaolinite would react with the large amount of sodium in the sea to produce an idealized sodium feldspar. No matter from which direction we approach the reaction, it is evident that the proportions of the different ions in solution will be controlled by the equilibrium constant. This constant would control the hydrogen ion concentration and hence the carbon dioxide pressure in the atmosphere.

Unfortunately, we have no evidence that such reactions actually do occur in sea water, so we do not know if their role in determining the pH of sea water is important.

Oxidation-Reduction in the Ocean

A piece of steel exposed to the atmosphere and rain undergoes oxidation, resulting in a product we call rust, which is composed of the minerals goethite (FeOOH) and possible hematite (Fe_2O_3). A simplified equation for this is:

$$2\ Fe^0 \text{ (metallic iron)} + \tfrac{3}{2}\ O_2 = Fe_2O_3 \text{ (hematite)}$$

This reaction can also be written as the sum of two half-reactions: one in which the iron is oxidized and the other in which the oxygen is reduced. Oxidation is the loss of electrons, reduction is the gain of electrons. Hence, an "oxidation-reduction reaction" actually involves a transfer of electrons.

Oxidation: $2\ Fe^0 \text{ (metallic iron)} = 2\ Fe^{+3} + 6 \text{ electrons}$

Reduction: $\tfrac{3}{2}\ O_2 + 6 \text{ electrons} = 3\ O^{-2}$

Sum of half reactions: $2\ Fe^0 + \tfrac{3}{2}\ O_2 = Fe_2O_3$

In our atmosphere the oxygen pressure is so large, and maintained thus as the result of photosynthesis, that the oxidation of metallic iron proceeds to completion. If we had a closed system, however, and a limited amount of available oxygen, the reaction would convert to iron oxide enough metallic iron to lower the oxygen pressure to an equilibrium value.

Since electrons are transferred in oxidation-reduction reactions, if we could imagine that the reaction takes place by the transfer of the electrons through a wire with a voltmeter connected across the wire, we would see a voltage potential difference between the two half-cells until the reaction had reached equilibrium. Hence, we can think of each half-cell as having a voltage potential. At equilibrium the oxidation potential would just equal the reduction potential. Of course, what we have been talking about has been put into practical use in the form of batteries. Equilibrium in a battery means a dead battery. The oxidation-reduction potential of a half-reaction relative to the oxidation-reduction of the idealized half-reaction (shown below) is called the Eh.

$$\frac{1}{2} H_2 \text{ (1 atm.)} \rightarrow H^+ \text{ (aqueous; thermodynamically equal to 1 mole per liter)} + \text{electron}$$

In our convention a positive Eh indicates that the half-reaction will couple with another half-cell and be reduced at the expense of the oxidation of the other species.

In normal sea water the half-reaction that determines the oxidation-reduction potential (Eh) of sea water is:

$$2\,H_2O = 4 \text{ electrons} + O_2 + 4\,H^+$$

All reactions in open sea water involving the transfer of electrons are controlled by this half-cell, and the Eh is fixed because the pressure of oxygen and the hydrogen ion concentration are determined for the ocean (Fig. 9-1).

Generally, ocean water is a highly oxidizing solution so that for many elements the higher oxidation states exist in solution. In nearshore areas and

FIG. 9-1 The relationship of pH, Eh, and the partial pressure of oxygen in the atmosphere. Although P_{O_2} is actually equal to 0.2 atmospheres instead of one atmosphere, the difference between the two lines is almost imperceptible. The relationship for P_{O_2} equal to 10^{-20} atmospheres is drawn for reference. The lines represent the half-cell: $2H_2O = 4$ electrons $+ O_2 + 4H^+$. At a pH $= 8$ and 0.2 atmosphere P_{O_2}, the Eh is about 0.75 volts.

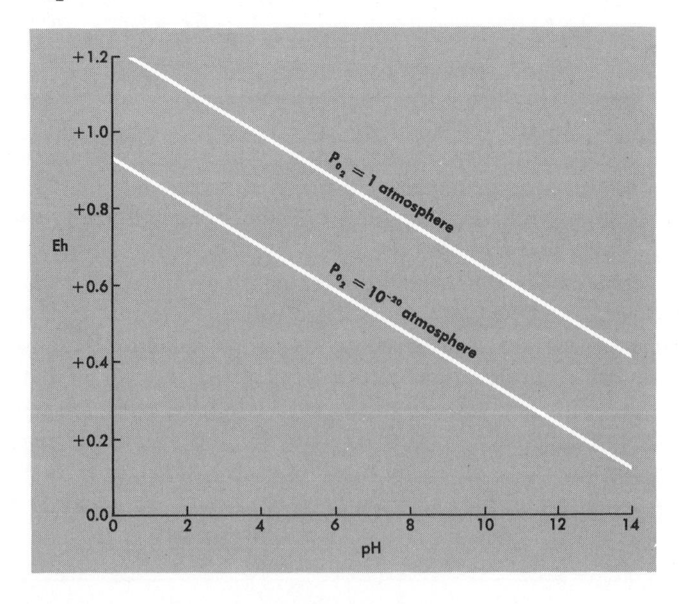

restricted basins, however, the effect of organisms in depleting the dissolved oxygen content is to lower the Eh and result in more reduced states of some of the elements.

THE BIOLOGICAL MODIFICATION OF OCEAN CHEMISTRY

The Nutrient Elements—Carbon, Oxygen, Nitrogen, and Phosphorus

Aside from water, the most prominent compounds in plants and animals are those containing carbon, oxygen, nitrogen, and phosphorus, in the form of amino acids (which make up proteins), fats, starches, sugars, and the phosphorus-containing compounds such as ATP (adenosine triphosphate), which are important for energy transfer within organisms. In the oceans, the constituent elements are available in solution as dissolved bicarbonate, phosphate, and nitrate.

The primary biological activity in the oceans, as on land, is the fixing of carbon by photosynthesis. All other life in the oceans depends on this first step. As we have seen, this process is accomplished primarily by single-celled organisms whose shells are commonly preserved in deep-sea sediments.

Carbon, nitrogen, and phosphorus are extracted from solution in the top hundred meters of the ocean where enough light penetrates for photosynthesis. The food chain then continues through a sequence of organisms living mainly in the surface layers, but some organic particles settle through the water. In the deeper waters bacterial action destroys much of this fine organic material and returns the nutrient elements to ionic form. The result of these events is that the surface waters are depleted in the dissolved nutrient elements and the deeper waters are enriched. If the nutrient elements were not returned to the surface waters by upwelling as described in Chapter 3, the concentration would soon become so low that the primary productivity would be diminished. At rates of biological productivity determined for the oceans, this depletion would lead to a virtually lifeless sea in less than a year.

The steady-state concentration gradients of phosphate and the other nutrient elements (Fig. 9-2) reflect the process of removal of dissolved nutrients from the surface waters, transport downward as particles, regeneration, and resupply by mixing. In areas of intense upwelling, water from depths to 300 meters is even more rapidly brought to the surface, thus supplying nutrient elements at a faster rate than that of normal eddy diffusion and small-scale convection. This upwelling results in greatly enhanced biological productivity.

The cycle of the nutrient elements is shown diagrammatically in Fig. 9-3. In biological processes oxygen plays an important role either as the by-product of photosynthesis or as a requirement for metabolism. The oxygen concentrations observed in sea water are close to those expected if equilibrium for gas

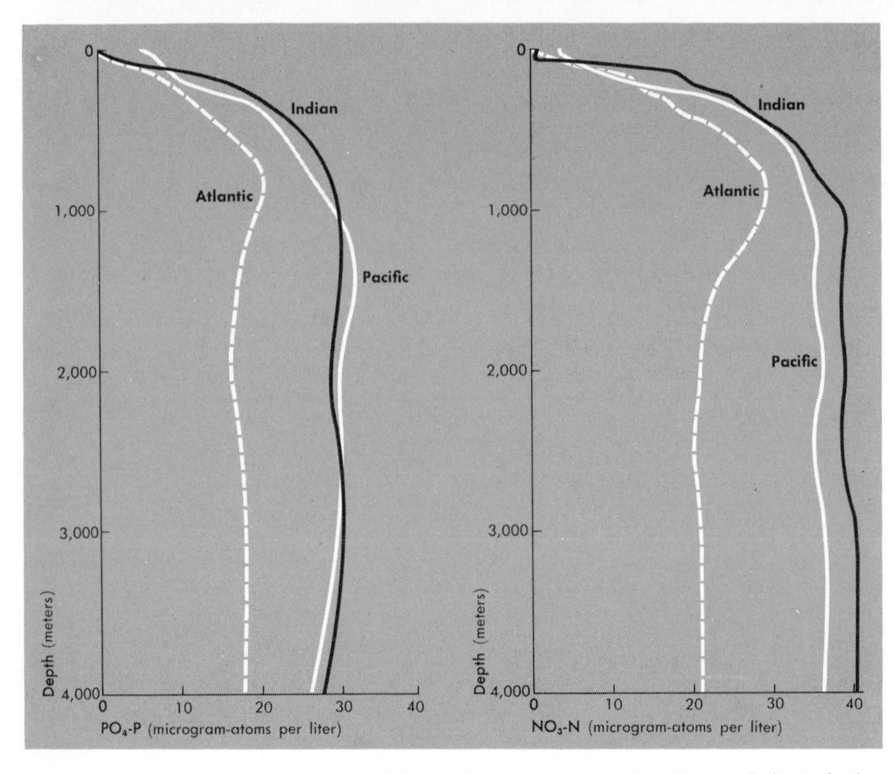

FIG. 9-2 (Above) Vertical distribution of the nutrient components, phosphate and nitrate, in the Atlantic, Pacific, and Indian Oceans. (After Sverdrup, Johnson, and Fleming, 1942.)

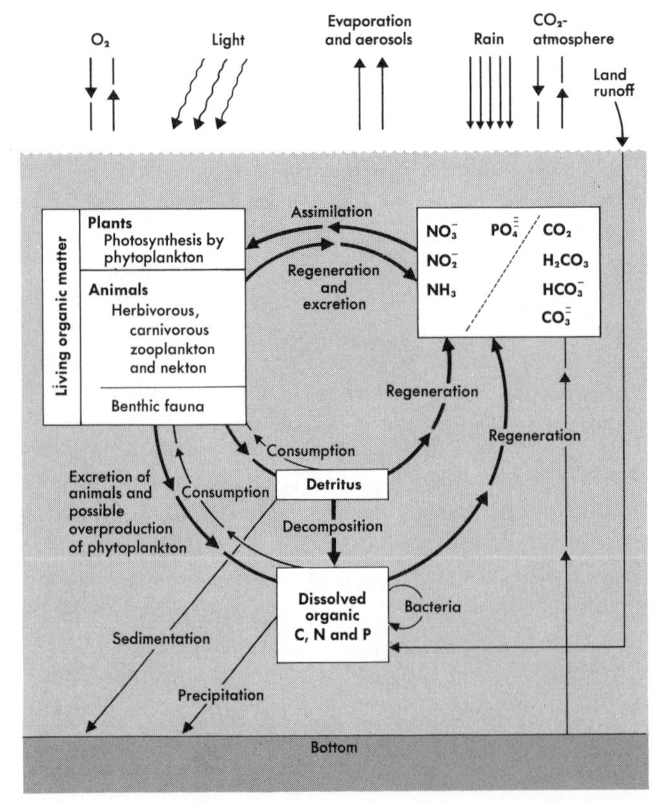

FIG. 9-3 (Left) A schematic diagram of the biogeochemical cycles of phosphorus, nitrogen, carbon, and oxygen. (After Duursma, 1961.)

distributions were established between surface ocean water and the atmosphere. For example, the cold Antarctic Bottom Water was formed in the Antarctic in contact with the atmosphere, hence should have a relatively high oxygen concentration. Yet values larger than the saturation value of oxygen are encountered in surface layers because of the production of oxygen by photosynthesis. Deeper, however, where photosynthesis is no longer possible, organic compounds are burned for food with the help of dissolved oxygen, thus causing a decrease in the dissolved oxygen concentration to values below saturation.

If aeration by advection, convection, or eddy diffusion is too slow, compared to the rate at which biologically useable organic material is supplied, the oxygen is rapidly used up and stagnation results at depth. Deep oceanic basins such as the Black Sea, the Cariaco Trench in the Caribbean, and fjords are examples of such areas. Under stagnant conditions the organic compounds are utilized by certain bacteria that obtain oxygen by reducing the plentiful sulfate ion ($SO_4^=$) in sea water to hydrogen sulfide (H_2S). This process also occurs in muds where aeration is inhibited by the higher rate of biodegradation than oxygen diffusion into the sediment.

The Composition of Plankton

In addition to the nutrient elements, other elements are included in the tissue of plankton. The exact mechanism of incorporation may vary. For example, cobalt is needed for the structure of vitamin B_{12} and copper in hemocyanin in the blood of crustacea, whereas some metals may be found in hard parts. There are other shell-forming elements besides the obvious calcium and silicon. Strontium as strontium sulfate tests of certain organisms is the most obvious, but others may be trapped in the lattices of all these major shell-forming elements. Thus, elements such as radium, magnesium, and barium make up part of the composition of shells. Finally, adsorption to chitinous and mucous tissue is an association of great importance. The compounds making up these substances are called polysaccharides and have metal sequestering bonds that tend to act as accumulators of metals from the medium.

Because of the high diversity in planktonic populations any one plankton tow may differ from another in both species composition and chemical composition. We can, however, divide the plankton into a *phytoplankton* fraction composed of microscopic photosynthesizing organisms like diatoms and coccolithophorids, as well as plant tissue without hard parts; a *zooplankton* fraction composed primarily of crustacea, larvae of various phyla, foraminifera, radiolaria, and a special radiolarialike group that deposits strontium sulfate tests. The phytoplankton are generally smaller in size than the zooplankton; thus the two groups can be discriminated by the mesh size of the nylon net used in towing.

Table 9-3 gives some average elemental concentrations of these two major groups of plankton. Obviously many metals are concentrated in plankton relative to sea water just as nitrogen and phosphorus are. Indeed, to a first approximation the trace elements are concentrated with the same efficiency as the nutrient elements.

Table 9-3 The Chemical Composition of Plankton (in Units of Micrograms of Element Per Gram of Dry Weight of Plankton)*

Element	Phytoplankton	Zooplankton
Si	58,000	—
Na	110,000	68,000
K	12,000	11,000
Mg	14,000	8,500
Ca	6,100	15,000
Sr	320	440
Ba	110	25
Al	200	23
Fe	650	96
Mn	9	4
Ti	≤ 30	—
Cr	≤ 4	—
Cu	8.5	14
Ni	4	6
Zn	54	120
Ag	0.4	0.1
Cd	2	2
Pb	8	2
Hg	0.2	0.1

* *From the data of Martin and Knauer (1973).*

What happens to the trace metals as the plankton are processed in the food chain? We know that phosphorus and nitrogen are regenerated as dissolved species within the top 500 meters and that very little of these elements reaches the ocean floor (see Chapter 6). Where phosphorus is high in deep-sea sediments many of the trace elements are also high, but this is not common for most of the ocean basin. We conclude that most of the elements tied up in the living tissue of plankton are returned to the sea water at the same rate as are phosphorus and nitrogen.

There are, however, the remains of planktonic organisms that do reach the ocean bottom. Pteropod tests, for example, contain 50 to 100 parts per million of iron and other trace elements. Probably foraminiferan tests and coccoliths also transport metals to the deep ocean floor. Two other major transporting agents are fecal pellets and chitinous tests of crustacea (copepods).

Estuarine Processes

Estuarine circulation driven by the supply of fresh water by rivers provides an environment and a mechanism for the modification of the composition of the ocean. The biological productivity is high in estuaries and the depth is shallower than the open ocean so that a large flux of organic material of planktonic origin reaches the bottom. There it joins the flux of material transported by stream and that derived from the margins of the basin by wave erosion and storm transport. This detrital material not only includes clays, silts, and sands, but it is also rich in organic material from soil profiles and marginal salt marshes.

The organic-rich deposits provide food for benthic organisms that live in and on the sediment. By the mechanisms described as *bioturbation* and *irrigation* (Fig. 9-4), large organisms such as worms, crustacea, and mollusks are able to metabolize the organic-rich sediment with oxygenated water from above. At greater depths and in parts of the sediment isolated from the aeration induced by the aerobic organisms, the sediment contains anoxic interstitial sea water. In such anoxic marine sediments sulfate-reducing bacteria convert the dissolved sulfate to hydrogen sulfide and metabolize the organic material by the general reaction:

$$\underset{\text{organic matter}}{2\,CH_2O} + SO_4^{=} = 2\,HCO_3^{-} + H_2S$$

FIG. 9-4 X-ray of a section of a box core obtained by diver at 14 meters depth in Long Island Sound. The sediment is heavily bioturbated. A storm prior to the raising of this core resuspended the top 1 to 2 centimeters of sediment which are stratified. Worm borings through the storm layers as well as the deeper layers indicate the initiation of bioturbation and irrigation of the sediment. Within days this sediment will be thoroughly bioturbated in the top 5 to 8 centimeters. (Courtesy of R. Aller.)

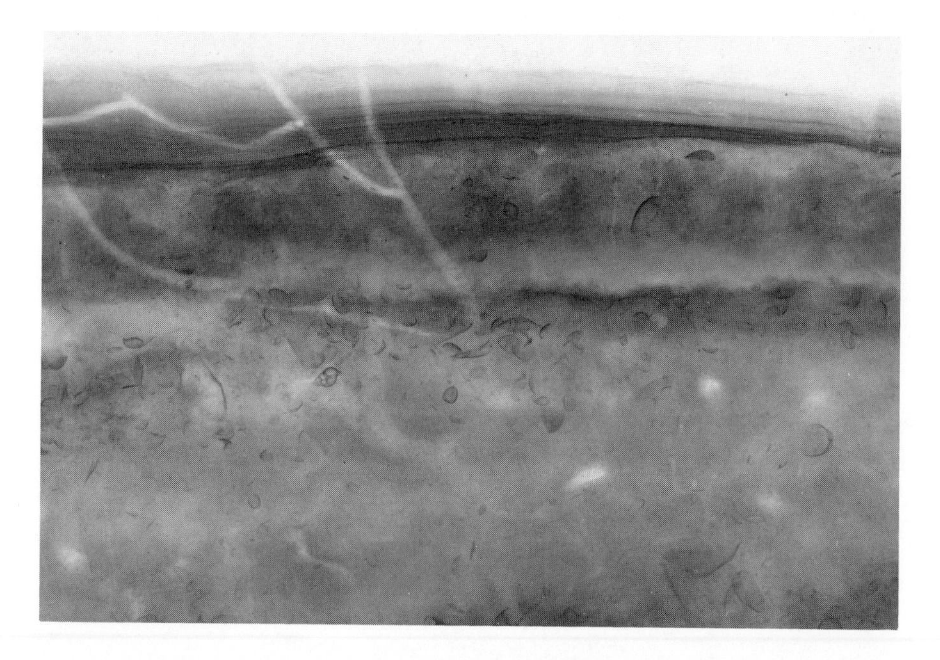

Hydrogen sulfide is a gas with the "rotten-egg" smell and the process can be certified by smelling an organic-rich mud flat at low tide. The presence of hydrogen sulfide results in an efficient trapping of many metals brought to the sediment by the various methods already described. At some depth in the sediment the sulfate is virtually all used up and the rate of transport of new sulfate from the overlying water is too slow to allow the process to continue. At these depths methane is produced by bacteria from the remaining organic material. By this process large amounts of potentially exploitable methane (or "natural gas") is produced.

Sometimes the biological productivity at the surface of the ocean delivers organic material to the depths at a faster rate than the utilized oxygen can be replenished in the deep-water column. Under these conditions the deep-water column itself goes anoxic and hydrogen sulfide becomes the dominant dissolved gas. The Black Sea and parts of the Gulf of California are good examples of this type of environment, as are many fjords. The Baltic Sea also becomes anoxic at times as a result of the interplay of productivity and exchange rate of water from the North Sea (Fig. 9-5).

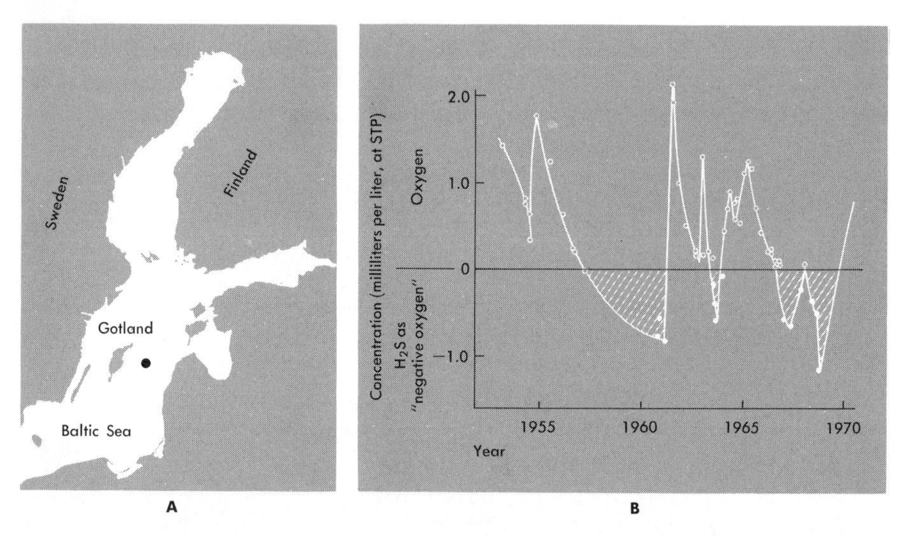

FIG. 9-5 (A) Map of the Baltic showing the location of the deep which is east of Gotland. (B) Oxygen and hydrogen sulfide at 200-meter depth in the Gotland Deep between 1953 and 1970. O_2 concentration is expressed in milliliters per liter ("STP" is standard temperature and pressure: 25°C and 1 atmosphere); H_2S is given as "negative oxygen" in the same units. (After Fonselius, 1969 and later data.)

The presence of hydrogen sulfide alters the chemistry of sea water drastically and the influence on the solubility of metals is marked (see Table 9-2). Whereas most metals are virtually immobilized in anoxic sediments, reduction of iron from Fe^{+3} to Fe^{+2} and manganese from Mn^{+4} to Mn^{+2} greatly increases their solubilities. Thus bioturbating organisms release to the overlying water a flux of these two reduced metallic ions. When they encounter the oxygenated water they are reoxidized and tend to precipitate as oxides. If some of this iron and manganese is transported to the deep sea as fine-grained particles they can influence the composition of deep-sea sediments.

It is possible to record the events of the last 100 years in the history of an estuary and in particular to study the impact of man on the area by using a dating technique based on one of the members of the uranium decay series, lead-210.

In Long Island Sound, a body of water surrounded by the high population and industrial areas of New York and Connecticut, the sediments accumulate at a rate of 0.45 centimeters per year (Fig. 9-6). This is about a thousand times faster than the average clay accumulation rate in the deep ocean. This rate can be assumed for most of the East Coast estuarine areas as a first approximation.

FIG. 9-6 A gravity core, 76 centimeters in length, was raised from (A) the center of Long Island Sound south of New Haven, Connecticut and (B) dated using excess Pb^{210} in the sediments. The source of the Pb^{210} is primarily the atmosphere and the dating technique is in principle the same as that used in Th^{230} dating of deep-sea cores. (From Thomson, Turekian, and McCaffrey, 1975.)

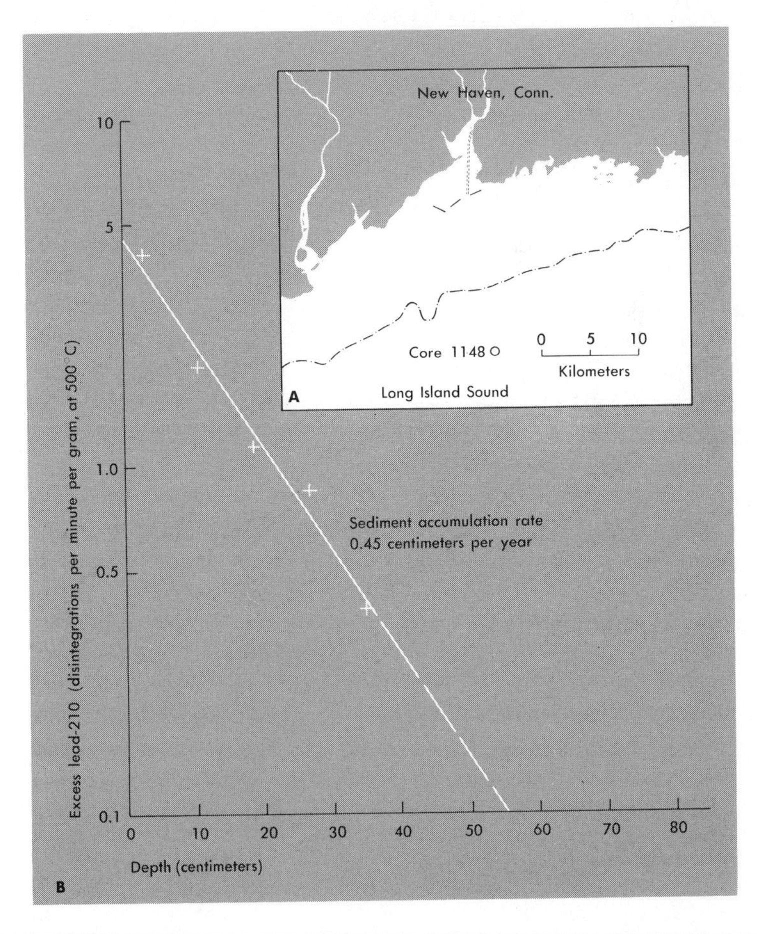

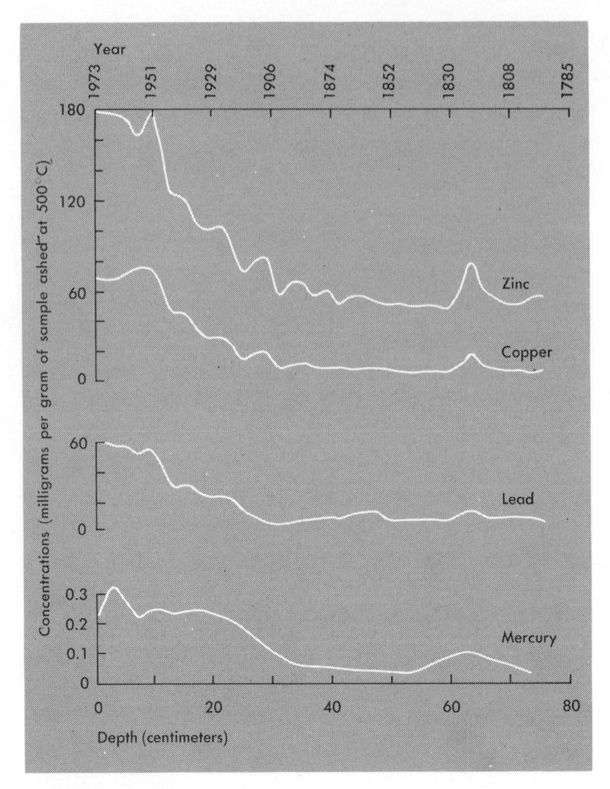

FIG. 9-7 The dated core of Fig. 9-6 was analyzed systematically down its length for the metals zinc, copper, lead, and mercury. The concentrations of all these elements have been increasing markedly since 1900. The primary source of these metals appears to be atmospherically transported industrial and urban pollutants. (From Thomson, Turekian, and McCaffrey, 1975.)

If we assume that the average depth of the semienclosed estuarine areas is about 40 meters, then all of these areas should be filled with sediment in less than 10,000 years. Indeed with the action of salt marsh building outward from the shore and the filling-in of these estuarine areas, we can expect a reshaping of the coast lines in the next 10,000 years. Other embayments may form to continue the chemical processes we have been discussing, but it is not certain how this will happen.

If we look at a dated core raised from these estuarine areas, we note that the more recent deposits are commonly richer in metals than are the older deposits (Fig. 9-7).This is a common finding in both coastal and lake sediments. The time of the increase in metals coincides with the rise of human occupation and industrialization of an area. Most of the anthropogenic metals appear to be transported to the body of water by atmospheric processes, although some are also transported via streams, sewer outfalls, and dumping barges.

Deep-Sea Processes

We noted above that marine plankton can transport metals, phosphorus, and nitrogen to depth. Except in relatively shallow nearshore waters and in areas of high productivity, in the open ocean (as discussed in Chapter 6), most of these elements are recycled through the upper water column via biological activity. Debris capable of transporting elements, of course, does reach the ocean bottom as seen in the calcareous and siliceous tests accumulating in deep-sea deposits. This biogenic debris also provides reactive surfaces on which further reactions can occur during their descent through the water column. Mucous coatings in fecal pellets and chitinous debris from a number of types of planktonic crustacea, for example, have strong capacities to sequester metal from the sea water through which they pass.

This indeed has been verified using a natural radioactive isotope of lead, lead-210, produced from dissolved radium-226 in the ocean. As lead-210 has a half-life of 22 years and the deep ocean water is one or two orders of magnitude older, we should expect lead-210 to be in equilibrium abundance with radium-226 (see Chapter 7). Any departure from the equilibrium ratio (the activities should be equal thus the activity ratio will be one at equilibrium) indicates that lead-210 is being removed by adsorption scavenging by particles falling through the water (much as thorium-230 is removed into the sediments and thereby used for dating deep-sea cores). It has been observed that, indeed, the higher the biological productivity in the overlying water, the greater the scavenging of lead-210 from the deep water column.

MATERIAL BALANCE

In Chapter 1 the concept of the hydrologic cycle was discussed. In this cycle water is transferred from the oceans to the continents and back again through the agencies of evaporation, precipitation, and stream runoff. In Chapter 8 the concept of a cycle involving the continents and the ocean floor was established through the model of plate tectonics. Mountains are formed, eroded, deposited into the sea, and returned in part as continental material along the convergences of the plates and in part returned to the mantle. These two determine the large-scale cycles of the elements in the crust of the Earth. The hydrologic cycle transports the material of the continents to the oceans and the tectonic cycle returns it, albeit in altered form.

We can construct a material balance for many of the major dissolved species carried by streams to the sea. The return cycle is presumed, although the time lags and pathways are known to be very complex.

Stream Supply Rates of Materials to the Ocean

The total annual supply of dissolved material to the ocean from the streams of the world is about 36×10^{14} grams. The supply rate of detritus is about 5 times as much. The composition of streams varies around the Earth because of differences in climate and types of local rock undergoing weathering; nevertheless the average concentration of the major ions in streams given in Table 9-4 is a reasonable starting point for our discussion.

Sodium, potassium, and silica (SiO_2) are derived from common silicate minerals such as the feldspars (potassium, sodium aluminosilicate) by the action of carbonic acid, which is itself derived from the action of bacteria on organic material.

Table 9-4 The Average Composition of Streams

	MEASURED CONCENTRATION		AMOUNT DUE TO MARINE AEROSOLS	DERIVED BY WEATHERING	
Component	Milligrams per liter	Millimoles per liter	Millimoles per liter	Millimoles per liter	Milligrams per liter
HCO_3^-	58.4	0.957	—	0.957	58.4
$SO_4^=$	11.2	0.116	0.011	0.105	10.1
Cl^-	7.8	0.219	0.219	0	0
NO_3^-	1.0	0.016	—	0.016	1.0
Ca^{++}	15.0	0.375	0.004	0.371	14.8
Mg^{++}	4.1	0.171	0.022	0.149	3.6
Na^+	6.3	0.274	0.188	0.086	2.0
K^+	2.3	0.059	0.004	0.055	2.1
SiO_2	13.1	0.211	—	0.211	13.1

Calcium and magnesium are derived primarily from the weathering of limestones and dolomites, although a small amount is supplied by the disintegration of silicate minerals. The main agent of decomposition again is carbonic acid. The reaction for calcium is written:

$$CaCO_3 + H_2O + CO_2 = Ca^{++} + 2\,HCO_3^-$$

Sulfate is formed by the oxidation of iron sulfide (pyrite), which is found as an important accessory mineral in many rock types, including sedimentary deposits. The oxidation of sulfide minerals to form sulfate ions produces hydrogen ions, which attack rocks.

In addition to these ions derived directly from the weathering of rocks, marine aerosols also supply ions to streams. In particular, this is seen as an increased sodium and chloride concentration. Indeed, most of the chloride in streams is of aerosol origin. The last column in Table 9-4 is the unrecycled burden of streams calculated by assigning all the chloride to the marine aerosol contribution.

The dissolved load of streams, as well as the sediments, brought to the oceans must ultimately be removed from the water if a steady-state composition is to be maintained.

The Marine Budget of Calcium

Calcium is removed from ocean water mainly by the deposition of calcium carbonate by marine organisms. Where does this take place? In nearshore deposits of molluskan shells and coral reefs (with their associated deposits) or in the deep-sea environment as coccoliths and foraminiferan shells? It is not possible to make an accurate survey of nearshore deposition rates because of the great variability of nearshore calcium carbonate fixation and sediment accumulation. It is possible, however, to determine the average accumulation rate of calcium carbonate in the deep oceans.

Table 9-5 Calcium Carbonate Balance in the Oceans

Ocean	Average Percentage of $CaCO_3$ in Deep-Sea Sediments	Average Clay Accumulation Rate (grams per square centimeter per 1,000 years)	Total $CaCO_3$ Accumulation Rate (10^{16} grams $CaCO_3$ per 1,000 years)
ATLANTIC OCEAN	43.7	1.20	77
PACIFIC OCEAN	37.8	0.30 ?	30
INDIAN OCEAN AND ALL OTHER SEAS			30
TOTAL			137
RATES OF ATLANTIC OCEAN AND ALL OTHER OCEANS EQUAL			154
RATE OF ALL OTHER OCEANS EQUAL TO HALF THE ATLANTIC OCEAN RATE			115
STREAM SUPPLY			122

In Chapter 7 we saw that rates of accumulation of clay and calcium carbonate components were determined in a large number of cores raised from the Atlantic Ocean based on radiocarbon dating and paleontologic correlation. Since most of these cores were, by design, rich in calcium carbonate, they represent a biased sampling of calcium carbonate rates. The correct method of determining the calcium carbonate accumulation rate in the Atlantic Ocean is to use the average clay accumulation rate and the average calcium carbonate concentration of the deep-sea sediments. The results of this type of estimation are presented in Table 9-5. It is doubtful that the total deep-sea accumulation rate of calcium carbonate in all the oceans other than the Atlantic is less than half that of the Atlantic or more than equal to it. In any case, the total rate is in the range of values calculated for calcium carbonate supply from the continents by streams. If much calcium carbonate is deposited on the continental shelves, the additional calcium may come from the alteration of deep-sea basalts.

Let us consider the long-term balance of calcium carbonate. At present rates of accumulation in the deep sea, it will take about 100 million years for all the calcium carbonate that has been deposited as limestones throughout geologic time and is now exposed above sea level on the continents to be removed and deposited in the deep sea. Table 9-5 shows that most of the calcium carbonate is apparently being deposited in the Atlantic Ocean. As the Atlantic is not bounded by convergence (or "subduction") zones, indeed the calcium carbonate will have to be retained in deep-sea sediments until the time that new spreading patterns destroy the Atlantic Ocean floor. In the Pacific calcium carbonate deposits are accumulating in the ridge areas and the high productivity region along the equator but these are already marked for recycling at the present plate boundaries. Eventually all of the oceanic sediment repository will be recycled and will return to the continents.

At times when the continents were more submerged than at present, the major repository of calcium carbonate would be as shallow water deposits. As we have seen, the oceans are supersaturated with respect to calcium carbonate at the surface and generally undersaturated at depth. Thus large inland (or "epeiric") seas would act as sinks for calcium carbonate just as the Bahamas platform acts as one today on a much smaller scale.

If, just prior to the Cambrian period 600 million years ago, before interior seas and continental shelf areas became extensive, the continents had been "washed clean" of calcium carbonate, and if the amount of calcium carbonate dissolved in the oceans was about the same then as now, the main area of deposition for several hundred million years before the beginning of the Cambrian period must have been the deep sea. At the onset of the Cambrian period, however, with the extensive development of shallow-water areas and the evolution therein of a large number of organisms that deposited calcium carbonate, the transfer of calcium carbonate from the deep-sea bottom to the shallow water areas progressed.

The Marine Budget of Silicon

There are two major ways in which silicon is removed from sea water: (1) by inorganic reactions with the clay minerals, and (2) by organic removal by diatoms, radiolaria, and other organisms.

Although the evidence is circumstantial, it appears that silicon is removed at least in part by reaction with clay minerals. The best argument proposed for this type of removal is that when clay minerals are put into a sea water solution containing silicon in concentration greater than that in the sea, the silicon concentration is decreased. This appears to be happening, in nature, along the Mississippi delta, where silica-rich, muddy Mississippi River water mixes with Gulf of Mexico water. This effect is not evident at the mouths of other large rivers, in part probably because the amount of clay minerals is low compared to the Mississippi.

The evidence for biological removal is direct, as we have seen in Chapter 5. Aside from the Antarctic, Arctic, and equatorial Pacific sites of intense biological accumulation of silicon, the major upwelling areas on the eastern boundaries of the oceans are major repositories. One of the most spectacular examples of such deposition is the Gulf of California, where the total rate of silicon deposition exceeds the supply from local streams by a factor of 100, indicating that it is being pumped in from the open ocean.

The major dissolved fluxes of both calcium and silicon are in the Atlantic Ocean north of 20 degrees south latitude where more than 60 percent of the world's stream water runs into the sea. As we have seen, most of the calcium is deposited as calcium carbonate in the Atlantic Ocean, although, because of the long residence time of calcium in the oceans, there is no direct relationship of area of supply and area of deposition. Although the mean residence time of silicon in the oceans is considerably shorter than it is for calcium (around 10^4 years for silicon compared to 10^6 years for calcium) the oceans circulate fast enough to separate supply from deposition points. Table 9-6 shows the material balance for silicon in the ocean and clearly indicates that most of the silicon supplied to the oceans is removed in the Pacific Ocean. The high latitudes are silicon repositories as is the east Equatorial Pacific, but the main repository must be in the upwelling areas in the eastern part of the Pacific Ocean.

Table 9-6 The Silicon Balance in the Oceans

Supply	10^{14} Grams SiO_2 per Year
STREAM SUPPLY RATE	4.3
SUBMARINE ALTERATION OF BASALTS	0.03
SOLUTION OF GLACIAL DEBRIS	?
TOTAL SUPPLY TO OCEANS	$\geq$ 4.33
Removal in Oceans	
ANTARCTIC	0.11–0.38
PACIFIC SUBARCTIC	0.2
EQUATORIAL PACIFIC	0.005
BERING SEA	0.10
OKHOTSK SEA	0.15
GULF OF CALIFORNIA	0.15
TOTAL IDENTIFIED REPOSITORIES	$\leq$ 0.99
SILICA REMOVED IN OTHER MARINE REPOSITORIES, PROBABLY IN UPWELLING AREAS PRIMARILY IN THE PACIFIC OCEAN	~ 3.34

The Marine Budgets of Sodium, Potassium, and Magnesium

These three elements are not deposited biogenically to any great extent. Magnesium can substitute for calcium in carbonates but no dominant pelagic calcium carbonate contains more than about 0.05 percent magnesium. The magnesium/calcium ratio deposited biogenically, then, is about 0.001, where as the ratio supplied by streams is 0.2. About 5 percent of the magnesium brought to the sea by streams is deposited with pelagic calcium carbonate and 95 percent must be accounted for in other repositories. The constraints on sodium and potassium are even more severe because calcium carbonate is an even smaller repository for these elements than it is for magnesium.

A highly probable method for removing the supplied sodium, magnesium, and potassium is by trapping these ions as part of the pore waters in marine sediments, both deep-sea and nearshore. If we assume that the detrital (plus potential sedimentary homogenic deposits of calcium carbonate and silicon oxide) load of rivers is 500 milligrams per liter, then we should expect to see associated with marine sediments in the pore waters the equivalent of 2 milligrams of sodium per liter, 2.1 milligrams of potassium per liter, and 3.6 milligrams of magnesium per liter. We can see if the balance works by using some reasonable numbers obtained from the study of marine sediments.

Assuming a sediment porosity of 50 percent would mean that for every cubic centimeter of solid sediment there would be a cubic centimeter of void space filled with sea water, and assuming a density of 2.4 grams per cubic centimeter for the solid and 1 gram per cubic centimeter for the sea water. The ratio of pore water sodium to detritus (plus shells) would then be about 0.004, the ratio delivered by streams. About 72 percent of the magnesium can be accounted for in this way, but only 36 percent of the potassium. There must then be another sink for magnesium and potassium, especially for the latter. Such a sink exists in basalts that form along the spreading ridge systems. The "freshest" basalts in these areas have low potassium concentrations, but the slightly altered ones take up potassium and, probably to a lesser extent, magnesium. This then is the most likely additional sink for these two elements.

As the oceanic sediments are subducted at the plate convergences the mixture of sediment and salt water reacts to produce metamorphic and igneous rocks typically seen in continental terranes. The chlorine is released in part as volcanic hydrogen chloride gas and in part as thermal waters like hot springs, geysers, and fumaroles.

The Marine Budget of Sulfur

Sulfur is brought to the oceans as the sulfate ion. Although it is utilized by deep-sea organisms in essential organic compounds, the concentration in sea water is sufficiently high to be unaffected by the normal oceanic biological cycles. In anoxic areas of the deep ocean, such as the deeper waters of the Black Sea discussed earlier, sulfate-reducing bacteria lower the sulfate concentration of sea water by producing hydrogen sulfide.

Sulfate reduction occurs to a larger degree in anoxic muds in shallow waters on the continental shelf where the organic content is especially high.

The hydrogen sulfide produced in the sediments does not commonly reach the overlying ocean water in great abundance because it reacts readily with iron oxide in the sediments to produce black iron sulfide. When tides are low or waves churn the sediments, hydrogen sulfide may escape and make its presence known by its characteristic rotten-egg smell.

Sulfate may also be removed from sea water by precipitation as calcium sulfate in areas of evaporation such as the Persian Gulf or the "salinas" of Mexico. Deposits of gypsum ($CaSO_4 \cdot 2H_2O$) are formed at low temperatures in these environments and evaporite formations are preserved.

appendix

CONVERSION FACTORS

Fathom (fm) = 6 feet (ft) = 1.829 meters (m)
Meter (m) = 100 centimeters (cm) = 39.37 inches (in) = 3.281 ft
Kilometer (km) = 0.6214 miles (mi)
Micron (μ) = 10^{-6} m = 10^{-4} cm
Centimeters per second (cm/sec) = 0.0360 km/hr = 0.0224 mi/hr
Liter (l) = 1000 cm^3 = 10^{-3} m^3 = 1.057 quarts
Kilogram (kg) = 10^3 grams (g) = 2.205 pounds (lb)
Microgram (μg) = 10^{-6} g
Year = 31,560,000 seconds
2.303 log (base 10)x = ln (base e or natural base)x

CONSTANTS

Avogadro's Number = 6.023×10^{23} g^{-1} $mole^{-1}$ (molecule per gram-molecular weight)
Universal Gravitational Constant = 6.670×10^{-8} dynes cm^2 g^{-2}
$\pi = 3.1416$
$e = 2.7182$

TERRESTRIAL CONSTANTS

Mass of the Earth = 5.976×10^{27} g
Area of the Earth = 510.100×10^6 km^2
Equatorial radius of the Earth = 6378.163 km
Polar radius of the Earth = 6356.177 km
Area of the oceans = 362.033×10^6 km^2
Mean depth of the oceans = 3729 m
Volume of the oceans = 1.350×10^9 km^3
Volume of the Earth = 1.083×10^{12} km^3
Standard acceleration of free fall on Earth = 980.665 cm sec^{-2}

suggestions for further reading and bibliography

CHAPTER 1

CLARK, S. P., JR., *Structure of the Earth.* Englewood Cliffs, N.J.: Prentice-Hall (1971).
TUREKIAN, K. K., *Chemistry of the Earth.* New York: Holt, Rinehart and Winston (1972).

CHAPTER 2

BURT, C. A. and C. L. DRAKE, eds., *The Geology of Continental Margins.* New York: Springer-Verlag (1974).
FISHER, R. L. and R. W. RAITT, *Deep-Sea Res., 9,* 423–443 (1962).
HEEZEN, B. C., M. THARP and M. EWING, *The Floors of the Oceans, I: The North Atlantic.* Boulder, Colo.: Geol. Soc. (1959).
HILL, M. N., *Physics and Chemistry of the Earth, 2,* 129–163. Elmsford, N.Y. Pergamon (1957).
LEYDEN, R., R. SHERIDAN and M. EWING, UNESCO-IUGS Symp. on Continental Drift Emphasizing the History of the South Atlantic Area, Montevideo, Uruguay. Oct. 1967, quoted in *The Earth's Crust and Upper Mantle,* Geophys. Monograph 13, P. J. Hart, ed. Washington, D.C.: Am. Geophysical Union (1969).
TALWANI, M., X. L. LEPICHON and M. EWING, *J. Geophys. Res., 70,* 341–352 (1965).

CHAPTER 3

DIETRICH, G., *General Oceanography.* New York: Wiley (1963).
MUNK, W. H., *Sci. Am., 193:34,* 96–102 (1955).
NEUMANN, G. and W. J. PIERSON, JR., *Principles of Physical Oceanography.* Englewood Cliffs, N.J.: Prentice-Hall (1966).
SPENCER, D. W., *Earth and Planet. Sci. Letters, 16,* 91–102 (1972).
STOMMEL, H., *Deep-Sea Res., 5,* 80–82 (1958).
SVERDRUP, H. V., M. W. JOHNSON and R. H. FLEMING, *The Oceans.* Englewood Cliffs, N.J.: Prentice-Hall (1942).

CHAPTER 4

BASCOM, W., *Waves and Beaches.* New York: Anchor Doubleday (1964).
DEFANT, A., *Physical Oceanography.* Oxford: Pergamon (1960).
GORDON, R. B., *Physics of the Earth.* New York: Holt, Rinehart and Winston (1972).
GROSS, M. G., *Oceanography.* Englewood Cliffs, N.J.: Prentice-Hall (1972).

CHAPTER 5

HEEZEN, B. C. and C. L. DRAKE, *Am. Assoc. Petro. Geol. Bull., 48,* 221–225 (1964).
HEEZEN, B. C. and C. D. HOLLISTER, *The Face of the Deep.* New York: Oxford (1971).
HEEZEN, B. C. and C. HOLLISTER, *Marine Geol. 1,* 141–174 (1964).
HILL, M. N., ed., *The Sea, 1, 2 and 3.* New York: Wiley (1962) Article by G. O. S. Arrhenius.
MASON, B., *Principles of Geochemistry.* New York: Wiley (1966).

CHAPTER 6

BERGER, W. H., *Marine Geol., 8,* 111–138 (1970).
BONATTI, E., *Trans. N.Y. Acad. Sci., 25,* 938–948 (1963).
BISCAYE, P. E., *Geol. Soc. Am. Bull., 76,* 803–832 (1965).
CALVERT, S. and N. B. PRICE, *Nature, 227,* 593–595 (1970).
GRIFFIN, J. J., H. WINDOM and E. D. GOLDBERG, *Deep-Sea Res., 15,* 433–459 (1968).
HORN, D. R., M. EWING, B. M. HORN and M. N. DELACH, *Marine Geol., 11,* 287–323 (1971).
MERO, J., *The Mineral Resources of the Sea.* Amsterdam: Elsevier (1965).
PETERSON, M. N., *Science, 154,* 1542–1544 (1966).
RILEY, J. P. and G. SKIRROW, ed. *Chemical Oceanography, 1 and 2.* London and New York: Academic Press (1965). Articles by F. Culkin and K. K. Turekian.
SEARS, M., ed., *Oceanography.* Washington, D.C.: American Association for the Advancement of Science (1961). Article by M. Bramlette.
SKORNYAKOVA, N. S. and P. F. ANDRUSCHENKO, *Lithology and Useful Minerals.* USSR: (Akad. Nauk), 21–36 (1964).
TUREKIAN, K. K. and J. IMBRIE, *Earth and Planet. Sci. Letters, 1,* 161–168 (1966).

CHAPTER 7

BROECKER, W. S. and J. VAN DONK, *Rev. Geophys. Space Phys.*, *8*, 169–198 (1970).
BROECKER, W. S., K. K. TUREKIAN and B. C. HEEZEN, *Am. J. Sci.*, *256*, 503–517 (1958).
COX, A., *Science*, *163*, 237–245 (1969).
DYMOND, J. R., *Science*, *152*, 1239–1241 (1966).
EMILIANI, C. and N. SHACKLETON, *Science*, *183*, 511–514 (1974).
SHACKLETON, N. and J. KENNETT, *Initial Reports of the Deep Sea Drilling Project*, *XXIX*, Washington, D.C.: U.S. Govt. Printing Office, 743–755 (1975).
SHACKLETON, N. and N. OPDYKE, *Quat. Res.*, *3*, 39–55 (1973).
TUREKIAN, K. K., ed., *Late Cenozoic Glacial Ages*. New Haven: Yale Univ. Press (1971). Article by J. Imbrie and N. Kipp.

CHAPTER 8

DEWEY, J. D., *Sci. Am.*, *226:16*, 56–66, (1972).
ISACKS, B., J. OLIVER and L. R. SYKES, *J. Geophys. Res.*, *73*, 5855–5899 (1968).
PHILLIPS, J. D., *Oceanus*, *XVII*, 21–27 (1973).
PITMAN, W. C. and J. R. HEIRTZLER, *Science*, *154*, 1164–1171 (1966).
PITMAN, W. C., R. L. LARSON and E. M. HERRON, *The Age of the Ocean Basins*. Boulder, Colo.: Geol. Soc. (1974).

CHAPTER 9

BERNER, R. A., *Principles of Chemical Sedimentology*. New York: McGraw-Hill (1971).
BROECKER, W. S., *Chemical Oceanography*. New York: Harcourt Brace Jovanovich (1974).
CRAIG, H., S. KRISHNASWAMI and B. L. K. SOMAYAJULU, *Earth and Planet. Sci. Letters*, *17*, 295–305 (1973).
DUURSMA, E. K., *Neth. J. Sea Res.*, *1*, 1–147 (1961).
FONSELIUS, S. H., *Hydrography of the Baltic Deep Basins III*, Report 23. Lund: Fishery Board of Sweden (1969).
GARRELS, R. M. and C. L. CHRIST, *Solutions, Minerals and Equilibria*. New York: Harper and Row (1965).
GARRELS, R. M. and F. T. MACKENZIE, *Evolution of Sedimentary Rocks*. New York: Norton (1971).
RILEY, J. P. and R. CHESTER, *Introduction to Marine Chemistry*. London and New York: Academic Press (1971).
THOMSON, J., K. K. TUREKIAN and R. J. McCAFFREY, in *Estuarine Research* (Ed. L. Cronin), *1*, 28–44 (1975).

index

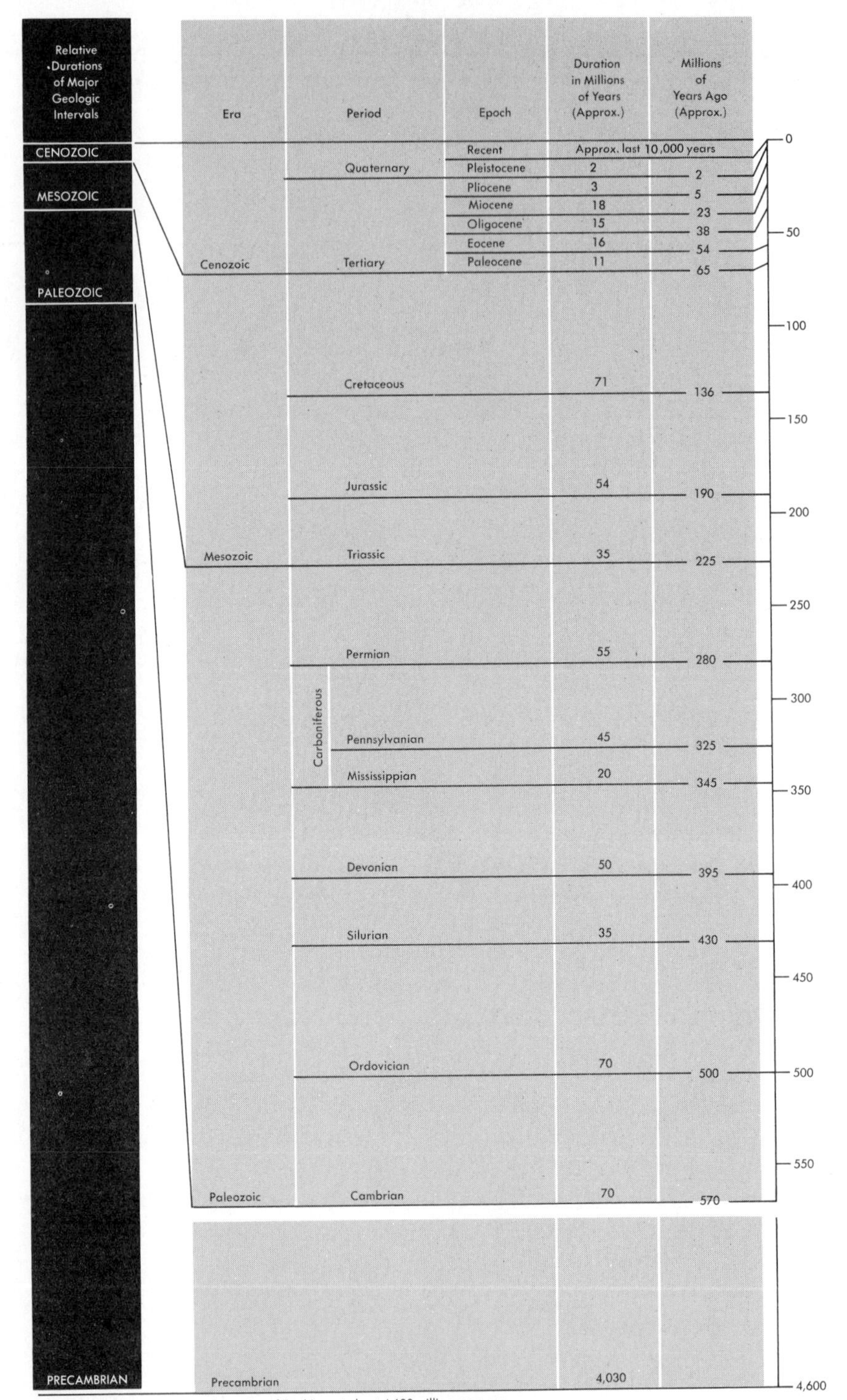

Relative Durations of Major Geologic Intervals
CENOZOIC
MESOZOIC
PALEOZOIC
PRECAMBRIAN
Era
Period
Epoch
Duration in Millions of Years (Approx.)
Millions of Years Ago (Approx.)
Recent
Approx. last 10,000 years
Quaternary
Pleistocene
2
2
Pliocene
3
5
Miocene
18
23
Oligocene
15
38
Eocene
16
54
Cenozoic
Tertiary
Paleocene
11
65
Cretaceous
71
136
Jurassic
54
190
Mesozoic
Triassic
35
225
Permian
55
280
Carboniferous
Pennsylvanian
45
325
Mississippian
20
345
Devonian
50
395
Silurian
35
430
Ordovician
70
500
Paleozoic
Cambrian
70
570
Precambrian
4,030
0
50
100
150
200
250
300
350
400
450
500
550
4,600
Millions of Years
Formation of Earth's crust about 4,600 million years ago